TRIAL BY FIRE

FOREST FIRES AND FORESTRY POLICY IN INDONESIA'S ERA OF CRISIS AND REFORM

CHARLES VICTOR BARBER
JAMES SCHWEITHELM

WORLD RESOURCES INSTITUTE
FOREST FRONTIERS INITIATIVE
IN COLLABORATION WITH
WWF WWF-INDONESIA & TELAPAK INDONESIA FOUNDATION

Carollyne Hutter
Senior Editor

Hyacinth Billings
Production Manager

Designed by:
Papyrus Design Group, Washington, DC

Each World Resources Institute report represents a timely, scholarly treatment of a subject of public concern.
WRI takes responsibility for choosing the study topics and guaranteeing its authors and researchers freedom of inquiry.
It also solicits and responds to the guidance of advisory panels and expert reviewers.
Unless otherwise stated, however, all the interpretation and findings set forth in WRI publications are those of the authors.

CONTENTS

Appendixes

Maps

List of Boxes

List of Tables

ACKNOWLEDGMENTS

We would like to extend our thanks to the many people and institutions who contributed to the development of this report.

At WRI, valuable comments, support, and guidance came from Cristina Balboa, LauraLee Dooley, Antonio La Viña, Anthony Janetos, Nels Johnson, Kenton Miller, Frances Seymour, and Nigel Sizer. At Telapak Indonesia, Abdon Nababan, A. Ruwindrijarto, Rezal Kusumaatmadja, and Wardiyono all provided crucial input and guidance. At WWF-Indonesia, we were greatly assisted by Agus Purnomo, Tim Jessup, Fernando Gonzalez, Martin Hardiono, Mariani Pangaribuan, and Carey Yeager.

The report greatly benefited from external reviews provided by Emily Harwell, Hira Jhamtani, Hariadi Kartodihardjo, Tahir Qadri, and Michael Ross.

Valuable information and ideas were also provided by Mubariq Ahmad, Longgena Ginting, Emy Hafild, Derek Holmes, Johannes Huljus, David McCauley, James Tarrant, and Tom Walton.

Finally, we would like to thank the production staff at WRI: Siobhan Murray for preparing the maps, Debbie Farmer, Carol Rosen, and Nancy Levine for their fine editing, and Hyacinth Billings for managing production.

C.V.B.
J.S.

FOREWORD

The forest and land fires that engulfed vast areas of Indonesia in 1997 and 1998 were an unprecedented human and ecological disaster. A prolonged dry season caused by the El Niño climatic phenomenon created the conditions for the conflagration. But the fires were mostly ignited deliberately by plantation companies and others eager to clear forest land as rapidly and cheaply as possible, no matter what the consequences. This was not a "natural" disaster.

Nearly 10 million hectares burned, including parts of 17 protected forest areas, shrouding many towns in darkness at noon and exposing some 20 million people across Southeast Asia to harmful smoke-borne pollutants for months on end. Economic damages from the resultant breakdown of transportation, destruction of crops and timber, precipitous decline in tourism, additional health care costs, and other impacts have been conservatively estimated to have totaled around $10 billion. The toll on Indonesia's rich forest biodiversity is unknown, but is thought to have been extremely high as well.

Disastrous as the Indonesian fires were, they were only one symptom of a far greater disaster—the systematic plunder and destruction of Southeast Asia's greatest rainforests over the past three decades under the rule of the avaricious and authoritarian "New Order" regime of former President Suharto. As this report details, the fires of 1997-1998 were the direct and inevitable outcome of forest and land-use policies and practices unleashed by the Suharto regime and perpetuated by a corrupt culture of "crony capitalism" that elevated personal profit over public interest, the environment, or the rule of law. Top Suharto regime officials and their business cronies treated Indonesia's forests as their personal property for more than 30 years, liquidating valuable timber through reckless and destructive logging practices, clear-cutting forests for oil palm and pulp plantations, and running roughshod over the interests of the millions of forest-dependent peoples living in traditional communities throughout the archipelago. As a result, according to a 1999 remote sensing study, Indonesia lost at least 1.5 million hectares of forest every year from 1986 to 1997. Total forest loss since the advent of the Suharto era in the mid-1960s is thought to be at least 40 million hectares—an area the size of Germany and the Netherlands combined.

The pillage of Indonesia's forests proceeded despite repeated warnings from a handful of courageous and public-spirited government ministers and officials who did their best to reform the forestry sector but found their efforts repeatedly stymied and their hands tied. Hundreds of millions of dollars in development aid was also spent on well-meaning "forestry policy reform" efforts during the Suharto era, but with little effect. As a recent World Bank assessment of the tens of millions of dollars it loaned to Indonesia for forestry and forestry-related projects since the mid-1980s concluded "the Bank has so far been unable to influence the rate of deforestation or the degradation of forests in Indonesia. Extremely weak governance has been the most debilitating problem in the sector. . .[and has]. . .resulted in corruption and illegal activity."

As Indonesia embarks on this new millennium, however, the prospects for meaningful forest policy reform have greatly improved because of the dramatic economic and political convulsions of the past two years. While the flames raged across Borneo and Sumatra in 1997 and 1998, smouldering political tensions and economic stresses also ignited. Long a star performer in the East Asian "economic miracle" of the 1980s and 1990s, Indonesia's economy came crashing to earth during late 1997. As the World Bank's 1998 report on Indonesia's economy concluded, "no country in recent history, let alone one the size of Indonesia, has ever suffered such a dramatic reversal of fortune." Stripped of the gloss of rapid economic growth, Suharto was forced from office in mid-1998 by a tidal wave of demands for *reformasi*, turning over the government to transitional President B.J. Habibie. In mid-1999, Indonesians voted in their first free election in four decades, and reformist President Abdurrahman Wahid assumed office in October 1999.

The stage is now set for a thorough house-cleaning in Indonesia's forestry sector, and government officials, academics, environmental activists, and a reinvigorated press are now debating proposals that would have been considered absurd—or even seditious—in the mid-1990s. But many interest groups and practices from the Suharto era are well-entrenched, and change has been slow in coming, as the renewed outbreak of fires in March 2000—deliberately set to clear land for plantations, as in 1997—illustrates.

Many long-suffering forest-dependent Indonesians are not willing to wait for the government to act. In March 2000, for example, the Indonesian media reported that some 50 logging concessions covering 10 million hectares had been forced to suspend operations because of conflicts, sometimes violent, with aggrieved local communities, many of whom had occupied concessions and thrown the loggers out. It seems virtually certain that sweeping changes in the relationships between local communities, logging and plantation companies, and the government are imminent. The key question is whether government forest policy will lead and smooth the way for these changes, or will be dragged along by popular action—which is likely to turn increasingly violent—at the grassroots.

Trial by Fire: Forest Fires and Forestry Policy in Indonesia's Era of Crisis and Reform, written by WRI's Charles Victor Barber and WWF-Indonesia's James Schweithelm, uses the 1997-1998 fires as the starting point for a detailed critique of forest policy in the Suharto era and an elaboration of the key reforms needed to both slow the loss of Indonesia's forests and prevent future fire disasters. The report first analyzes the 1997-1998 fire disaster, reviewing the fires' impacts, costs, and causes. The authors go on to place the fires within the larger context of the destructive forest and land-use practices and policies that have characterized Indonesia for the past three decades. It is futile, they convincingly demonstrate, to believe that the recurrent and increasingly severe Indonesia fire problem can be solved in isolation from more general reforms of the forestry sector and other sectors that affect the use of forest lands and resources. The specific reforms that they recommend, therefore, not only serve the goal of ensuring that Southeast Asia is not periodically shrouded in a choking haze, but also support forest biodiversity conservation, sustainable forest-based fiber production, and recognition of the rights and interests of forest-dwelling peoples.

This report is the latest in a series produced by WRI's Forest Frontiers Initiative (FFI), a five-year, multidisciplinary effort to promote stewardship in and around the world's last major frontier forests by influencing investment, policy, and public opinion. It has been researched and written in collaboration with WWF-Indonesia and the Telapak Indonesia Foundation, two of the most active and respected Indonesian nongovernmental organizations working on forest policy reform. WRI is pleased to acknowledge the German Federal Ministry for Economic Cooperation and Development (BMZ), and AVINA for their support for this project.

Jonathan Lash
President
World Resources Institute

PART I

THE 1997-98 FOREST FIRES IN INDONESIA:

IMPACTS, COSTS, AND CAUSES

I. AN INFERNO IN WAITING: INDONESIA'S FOREST POLICY

Indonesia received unaccustomed attention in the world's headlines during the latter part of 1997 as forest and land fires raged in Kalimantan (Indonesian Borneo) and Sumatra. The fires pumped enough smoke into the air to blanket the entire region in haze, reaching as far north as southern Thailand and the Philippines, with Malaysia and Singapore being particularly affected. The fires burned out of control again in early 1998, the second year of perhaps the worst El Niño-related drought ever recorded in Indonesia.

Fires broke out once again in July-September 1999 in parts of Sumatra and Kalimantan, with satellites detecting over 500 "hot spots" in Sumatra's Riau province alone over one week in late July. Haze from the fires obstructed air transport, forced school closures, and raised pollution to hazardous levels.[1] By mid-September, the sun was completely obscured for days in parts of southern Kalimantan, hampering daily activities and causing a rise of respiratory-related medical complaints.[2] The 1999 fires were not of the same magnitude as those of 1997-98, but they raised fears that another prolonged dry season, expected by some experts for 2000, could soon lead to a repeat of the 1997-98 disaster.[3]

The health and economic effects of the 1997-98 fires and haze on Indonesia and surrounding countries, in addition to the enormous impacts on tropical forest ecosystems, biodiversity, and the Earth's atmosphere, prompted some observers to label the fires a global natural disaster. Two elements of the catastrophe were of particular concern to many observers. First, preliminary evidence indicated that most of the fires were set intentionally by timber and agribusiness firms intent on clearing land as cheaply as possible. Second, the Indonesian government's response to the disaster was perceived as generally weak, uncoordinated, and defensive, with the exception of the Ministry of Environment's monitoring and public information efforts and its forthright identification of the role of big business in setting many of the fires.

Disastrous as they were, these fires, and the weak government response, are only a symptom of long-standing forestry and land-use policies and practices that have degraded and deforested vast areas of Indonesia and brought hardship to millions of indigenous and local people while enriching a small group in the ruling circle. With some 75 percent of the nation's land area legally designated as forestland, these policies and practices have had profound effects on the nation as a whole, the forest fires being only the most recent and visible.

Indonesian forest policies have provided powerful legal incentives for "cut-and-run" resource extraction and have failed to create effective mechanisms for enforcing even minimum standards of forest resource stewardship.

SUHARTO'S REGIME AND THE FATE OF THE FOREST

Indonesia is currently in transition from the iron-fisted 32-year rule of President Suharto, whose "New Order" regime ended with his resignation in May 1998. In the forest and natural resources sector, the New Order political economy was characterized by a heavily centralized bureaucracy and industry, effectively dominated by a small number of corporate conglomerates with close connections to top politicians. These business groups and their bureaucratic cronies were essentially above the law for three decades, seeking short-term profits at the expense of the environment and local communities while enjoying the protection of a legal and political system in which neither industry nor the bureaucracy could be held accountable.

Indonesian forest policies have provided powerful legal incentives for "cut-and-run" resource extraction and have failed to create effective mechanisms for enforcing even minimum standards of forest resource stewardship. In addition, Indonesia has regularly been rated by businessmen as one of the most corrupt countries on Earth, where bribery and payoffs are an entrenched way of life.[4] Human rights abuse has been widespread (and has been frequently linked to conflicts between local people and elites over natural resources). Local police and military units have often served as a kind of private army to repress popular resistance to the exploitation of forests and other natural resources.

Reinforcing this forest exploitation system was an official ideology that excluded local and indigenous communities from access to forest lands and resources while at the same time using them as scapegoats for the negative consequences of government policy and private sector behavior—such as forest fires. Although Suharto was forced from office in May 1998, much of his heavily entrenched system and the elite that profited from it remain.

INDONESIA'S FOREST RESOURCES

Although some 143 million hectares (ha)—nearly three-fourths of Indonesia's land area—are legally classified as "forestland" of various types, estimates from the early 1990s of actual forest cover range from 92.4 million to 113 million ha.[1] A 1997 World Resources Institute (WRI) analysis warned, however, that only about 53 million ha of "frontier forest"–relatively undisturbed areas of forest large enough to maintain all of their biodiversity–remain in Indonesia.[2] Although various sources had estimated the country's annual deforestation rate at between 0.6 million and 1.2 million ha,[3] a mapping effort carried out with support from the World Bank during 1999 concluded that the average annual deforestation rate since 1986 has actually been about 1.5 million ha, much of it "caused by forest fires, often ignited by people clearing land cheaply for plantations."[4] Some 30 percent of Sumatra's forest cover vanished during this period, according to the World Bank study. (See the Table below.) It is probable, therefore, that the WRI figure is closer to the truth than earlier estimates. If current trends continue, virtually all nonswampy lowland forests in Kalimantan and Sumatra will be destroyed by 2010.[5]

Biologically, these forests are extremely diverse. Although Indonesia occupies only 1.3 percent of the world's land area, it possesses about 10 percent of the world's flowering plant species, 12 percent of all mammal species, 17 percent of all reptile and amphibian species, and 17 percent of all bird species.[6] The lowland forests of Sumatra and Kalimantan are among the most species-rich on Earth, and there is a wide range of other forest types, each with its own flora and fauna.

In addition to acting as a storehouse of biological riches, Indonesia's forests yield products that have helped develop the nation's economy and provide income for millions of people living in and around forests. At the end of 1995, 585 logging concessions held 20-year rights to cut timber from approximately 62 million ha,[7] producing some $5.5 billion in annual export revenues (15 percent of the national total), in addition to supplying the large domestic market.[8] In October 1998, the chair of the Indonesian Forestry Society reported that there were 421 logging firms, 1,701 sawmill companies, 115 plywood firms, and 6 pulp and paper companies. In mid-1999, the area covered by active concessions had decreased to 51.5 million ha.[9] The report noted that plywood exports in 1997 totaled 7.85 million cubic meters (m³), or 80 percent of total Indonesian plywood production, and were worth $3.58 billion, making Indonesia the world's biggest plywood producer in that year.[10] In 1996, total output from forest-related activities was about $20 billion, or about 10 percent of gross domestic product (GDP). Forest-related employment amounted to about 800,000 in the formal sector and many more in the nonformal sector, and royalties and other government revenues from forest operations exceeded $1 billion per year.[11]

Indonesia's forests yield many nontimber forest products, the most valuable of which are rattan canes, which had an export value of $360 million in 1994.[12] The forests also provide valuable environmental services such as protecting the hydrological balance of watersheds and storing carbon that would otherwise increase the concentration of greenhouse gases in the Earth's atmosphere.

Indonesia's forests are home to a large but undetermined number of forest-dwelling or forest-dependent communities. Estimates of the precise number of these communities vary wildly—from 1.5 million to 65 million people, depending on which definitions are used and which policy agenda is at stake.[13] Many of these forest dwellers live by long-sustainable "portfolio" economic strategies that combine shifting cultivation of rice and other food crops with fishing, hunting, gathering forest products such as rattan, honey, and resins for use and sale, and cultivating tree crops such as rubber. Many of these local values of the forest are poorly appreciated, however, because they are not reflected in formal market transactions.

Notes:
1. GOI, 1991: 9.
2. Bryant, Neilsen, and Tangley, 1997: 21.
3. Sunderlin and Resosudarmo, 1996.
4. World Bank, 1999c. The 1999 mapping exercise upon which the World Bank based its new deforestation estimates was carried out by the mapping and inventory division of the Ministry of Forestry and Estate Crops as one of the conditions required by the World Bank for its Policy Support Reform Loan II to the government. The mapping was done at reconnaissance level only based on interpretation from digital Landsat imagery at a scale of 1:500,000, without field checks, and therefore must be regarded as provisional. The project used imagery from 1996 or later wherever available, although in some areas imagery from 1994 and 1995 had to be used. The methods used do not permit an analysis of the quality of forest cover, only current distribution and regions and rates of removal.
5. Ibid.
6. BAPPENAS, 1993.
7. Brown, 1999.
8. Sunderlin and Resosudarmo, 1996.
9. World Bank, 1999c.
10. "Many Timber Firms Facing Closure in Indonesia," Cable News Network, October 7, 1998.
11. World Bank, 1999a.
12. De Beer and McDermott, 1996: 74.
13. Zerner, 1992: 4.

Deforestation in Sumatra and Kalimantan, 1985–97

	1985		1997		DEFORESTATION		
	FOREST	**PERCENT OF TOTAL AREA**	**FOREST**	**PERCENT OF TOTAL AREA**	**DECREASE 1985-1997**	**PERCENT LOSS**	**HECTARES PER YEAR**
SUMATRA	23,324,000	49	16,632,000	35	6,691,000	29	558,000
KALIMANTAN	39,986,000	75	31,029,000	59	8,957,000	22	746,000
TOTAL	63,310,000	63	47,661,000	47	15,648,000	26	1,304,000

Source: World Bank, 1999c.
Note: These figures should be taken as provisional and subject to future revision following further analysis of satellite data and ground-checking being carried out in 1999-2000.

INDONESIA'S ECONOMIC AND POLITICAL CRISES

The economic and political crises that have wracked Indonesia since mid-1997 have transformed the political and economic context for natural resource management. They have profound implications for efforts to prevent future forest fires and for broader forest policy reforms.

The East Asian economic crisis that began with the devaluation of the Thai baht in July 1997 affected Indonesia more severely than any other country in the region. By July 1998 the value of its currency had fallen by 80 percent, inflation had risen to more than 50 percent, and urban unemployment had soared to unprecedented levels. The Food and Agriculture Organization of the United Nations (FAO) estimated in that month that about 40 million people, or 20 percent of the population, were vulnerable to food scarcity. The government reported that 40 percent of the population was living below the official poverty line, up from only 11 percent in 1996. Coming after nearly three decades of uninterrupted rapid economic growth, these developments represent a unique historical debacle. In the words of the World Bank's mid-1998 assessment,

"No country in recent history, let alone one the size of Indonesia, has ever suffered such a dramatic reversal of fortune. The next years will be difficult and uncertain." The economy is expected to contract this year [1998] by 10–15 percent, inflation could exceed 80 percent, and the number of poor could well double.[1]"

While political unrest had been growing throughout 1996 and 1997, the economic crisis was a major catalyst for the crescendo of opposition and violence that drove President Suharto from office in May 1998. The legitimacy of his 32-year authoritarian rule was largely dependent on the delivery of continued economic growth, in exchange for which many elements of society were willing to tolerate rampant corruption, regular abuses of human rights, and the absence of democratic political processes. With the economy spiraling into depression and prices of basic foodstuffs and other commodities skyrocketing, support for the aging president—even among many who had long served in his government and military—evaporated in an explosion of student-led protests and violent riots.[2]

The demise of the New Order regime left Indonesia in a state of political limbo under the transitional government of Suharto protégé President B. J. Habibie. Parliamentary elections—the first rela-

tively free and fair elections in 44 years—were held in June 1999. In October 1999, the People's Consultative Assembly[3] elected Abdurrahman Wahid president, and Megawati Sukarnoputri vice-president.[4] Wahid lost no time in appointing a new cabinet reflecting the broad range of political parties that contested the June elections, elements of Suharto's old administrations, and representation from the armed forces. Hailed by the press as a "national unity" government, it is unclear at this writing how the change in government may affect forest policy.[5]

Virtually all elements of the political spectrum have adopted the rhetoric of *reformasi*—democratization of politics, respect for human rights, and the elimination of "corruption, collusion, and nepotism" (KKN). But *reformasi* means very different things to different people. For the many holdovers from the old regime who are still in power or are biding their time, it means removing the rough edges and the most blatant corruption from the current system but generally continuing business as usual. For students and other more radical reformers, it means nothing less than the complete burial of the New Order regime and the creation of a democratic political system. In Aceh and Irian Jaya—restive provinces with long-standing separatist movements—many view *reformasi* as an opportunity to gain at least a greater measure of autonomy, if not independence, from the central government. And in virtually all provinces, *reformasi* is equated with greater decentralization of political power and increased local access to the profits of natural resource exploitation.

The economic crisis and political upheavals have been accompanied by an increasing crescendo of civil violence. Much of it is directed against Indonesia's ethnic Chinese minority, long perceived by many Indonesians as unfairly dominating the economy and benefiting from the largesse of the Suharto regime.[6] In mid-May 1997, just before Suharto's resignation, riots destroyed Jakarta's Chinatown, left some 1,200 people dead, and were accompanied by horrifying rapes and other atrocities directed at the Chinese. Other incidents of anti-Chinese violence have taken place since then in a number of cities across the country. Christian-Muslim violence in the eastern province of Maluku left hundreds dead in the first part of 1999 and flared up again in July and October 1999. In March 1999, hundreds of migrant settlers from Madura, an island close to Java, were killed by local ethnic groups in West Kalimantan.

Generalized looting, arson, and theft have increased dramatically throughout the country, sometimes driven by the desperation of a populace sinking into poverty and hunger, but sometimes carried out by well-organized gangs taking advantage of the general chaos.[7] Strikes and protests over the price of food and basic necessities have become increasingly common throughout the country.

Just as law and order are breaking down in many areas, the legitimacy of the armed forces (ABRI), long a powerful political player, is at an all-time low, following revelations of its role in kidnapping and torturing democracy activists, the uncovering of mass graves of ABRI victims in the province of Aceh (where a long-simmering separatist rebellion was brutally put down in the early 1990s),[8] and the mid-1999 debacle in East Timor.[9]

Adding to Indonesia's woes, the 1997–98 drought caused by the periodic El Niño climatic phenomenon was perhaps the worst the country had experienced in some 50 years. Rice production dropped drastically as a result,[10] just as millions of newly unemployed urban workers were returning to their villages in search of livelihood. The drought also set the stage for the forest fires, although it was just one of many aggravating factors.

Notes:

1. World Bank, 1998a .

2. For accounts of the growing political tensions during 1996 and 1997, see Forrester and May 1999; for detailed accounts of the fall of Suharto, see Forrester and May 1999; "Indonesia's May Revolution," *Far Eastern Economic Review*, May 28, 1998; "Indonesia after Suharto," *Far Eastern Economic Review*, June 4, 1998; and M. Scott, "Indonesia Reborn?" *New York Review of Books*, August 13, 1998.

3. The People's Consultative Assembly (Majelis Permusyawaratan Rakyat or MPR) is composed of the elected members of the parliament, formally known as the People's Representative Assembly (Dewan Perwakilan Rakyat or DPR), and a number of members appointed by the government to represent various groups in society. Under Indonesia's constitution, the MPR elects the president and vice-president every 5 years, and lays down "Broad Guidelines of State Policy" for that period.

4. Abdurrahman Wahid, commonly called Gus Dur in Indonesia, is a widely respected moderate Islamic leader, while Megawati, the daughter of founding president Sukarno, was a key leader and focal point of opposition to the Suharto regime in the mid-1990s. Her party polled the most votes in the June parliamentary elections and was widely expected to win the presidential race.

President Habibie withdrew from the race just days before the vote, in large part because Armed Forces Chief Wiranto declined to stand as vice-president. As a result, Habibie's party, Golkar, did not field a candidate at all, leaving a two-way race between Gus Dur and Megawati. Gus Dur, a legendary political deal-maker, made deals with both the military and Golkar sufficient to ensure his election. Megawati, whose distaste for political horse-trading is well known, apparently expected her immense popularity with rank-and-file voters to carry the day in the presidential poll. Her supporters rioted throughout the country the night of her defeat, but appear to have been mollified by her election as vice-president and the warm embrace she has received from Gus Dur as his "full-partner" in overcoming Indonesia's economic and political troubles.

5. Nur Mahmudi Ismail, the new Minister of Forestry and Plantations installed in late October 1999, told the press in his first interview only that "I will consolidate the personnel of the ministry, strengthen morale and attitudes and secure their commitment to manage the country's natural resources in the public interest." ("New Ministers Look Ahead," *Jakarta Post*, October 27, 1999.) Trained in agriculture in the United States, Ismail is the co-founder and chairman of the Justice Party, one of the new reform-oriented parties established during 1998.

6. "Turning Point: Indonesia's Chinese Face a Hard Choice—Stand Up for Their Rights or Seek a New Life Abroad," *Far Eastern Economic Review*, July 30, 1998.

7. "Police Say Indonesia Faces Rising Tide of Unrest," *Reuters*, September 25, 1998; "Lawlessness Spreads as Looters Defy Army," *Straits Times* (Singapore), July 20, 1998.

8. "Indonesia: The Disappearing Army," *Sydney Morning Herald*, August 15, 1998.

9. A referendum on independence for this small ex-colony of Portugal, occupied by Indonesia since 1975, gave rise to a spasm of violence against independence supporters perpetrated by the Indonesian military through the "militia" gangs it armed and trained. At present writing, Indonesia has formally renounced its claim on East Timor, troops have left, and a three-year UN Transitional Authority for East Timor (UNTAET) has been established. The East Timor debacle was both a humiliation for the armed forces—which had vowed for decades never to give up the territory—and a major international exposure of its casual and systematic use of terror, arson, and murder to enforce its vision of "internal security."

10. In mid-1998, the Indonesian Central Bureau of Statistics estimated rice harvest failure resulting from the drought at 13 percent. An agriculture climatologist at the Bogor Agricultural Institute reported, however, that his research indicated a shortfall of 40 percent as a result of the drought ("'Horrific' Rice Forecast," *Reuters*, July 25, 1998).

Most official reaction to the 1997–98 fires by the Indonesian government and the international community has been technical and managerial in nature, focusing on improving drought-predicting technologies, strengthening administrative coordination, acquiring modern firefighting equipment, and the like. While these measures are certainly necessary, they will have only marginal effects unless the underlying political economy of forest resource use and management is significantly restructured.

Indeed, the focus on technical and bureaucratic responses to the forest fires may become an obstacle to meaningful action, to the extent that it turns the debate away from the broader forest policy issues that lie behind them. The official response to the fires mirrors decades of government efforts to justify and maintain the status quo and the concurrent promotion by donor agencies of incremental and technocratic "forest sector reforms" in a fashion palatable to the Suharto regime and its favored clients in the forest and agribusiness industries.

Until recently, one could justify these incremental, technocratic, and depoliticized approaches as the best to be expected, from either the government or donors, in the atmosphere of the New Order. The economy was growing rapidly, Indonesia was repaying its debts, and opposition to New Order policies was disorganized, muted, and rapidly silenced by Suharto's strong and stable regime. But all that has changed.

The fires dramatically announced to the world that something was seriously wrong in Indonesia's forests. The fate of these forests, the third largest tract of tropical forests on the planet and the largest in Asia, has always been of international concern. (*See Box 1 and Map 1.*) But the effects of the 1997–98 fires on neighboring countries made it clear that bad forest management in Southeast Asia's largest nation was an issue on which the international community might demand—not just suggest—reforms. At the same time, the country's economic collapse and the resulting need for massive international assistance have increased international leverage for pressing for forest policy reforms.[5] Finally, Suharto's departure has transformed the political landscape and created, at least for the time being, an unprecedented window of opportunity for domestic critics to influence forest policies. (*See Box 2.*)

FOREST FIRES: THE POLICY CONNECTION

This report argues that the fires of 1997 and 1998 were just the latest symptom of a destructive system of forest resource management carried out by the Suharto regime over 30 years. If the government —and the international donor community on which it so greatly depends in this period of economic crisis and recovery—are serious about preventing future infernos, the solution lies not so much in strengthening technical capacities for fire prediction, prevention, and mitigation as in a major restructuring of relationships between the state, the private sector, and the millions of forest-dependent peoples living in and on the fringes of the nation's forestlands.

The fires of 1997 and 1998 were just the latest symptom of a destructive system of forest resource management carried out by the Suharto regime over 30 years.

The reform agendas for reducing the threat of fire and for transforming forest policy from a catalyst for forest destruction into a guarantor of forest sustainability overlap to a great extent, and the first agenda cannot succeed without substantial progress on the second. This report therefore addresses both the immediate causes and the impacts of the 1997–98 fires and the broader question of how Indonesian forestry and forest land-use policies have contributed to forest degradation and deforestation processes—forest fires being an important element of these processes.

Although the forest and land fires (simply termed "forest fires" in this paper) of 1997–98 were unprecedented in their scale and effects, fire has been a regular feature of Indonesia's forest ecosystems for as long as humans have inhabited the archipelago.[6] The nation has a tropical climate with an annual pattern of wet and dry seasons—the result of monsoon winds that alternate seasonally between westerly and easterly directions. The westerly monsoon usually brings heavy rains to the western portion of Indonesia from September to April; drier easterly winds blow the rest of the year. Every few years this pattern is disrupted by the El Niño phenomenon (*see Box 3*), resulting in a prolonged dry season that, during severe El Niño episodes, may extend into the next regular dry season.

Normal "dry" seasons are actually relatively wet. Monthly dry season rainfall on the islands of Sumatra and Borneo usually exceeds 10 centimeters (cm), providing the moisture needed to sustain the lush evergreen tropical rainforests. During severe El Niño events, however, months often pass with no appreciable rainfall. Spells of one to two weeks with no rain create conditions dry enough for intentional burning of degraded forest and brush areas, but mature intact rainforests will burn only after considerably longer dry periods.

A Recurring Phenomenon

Despite the common puzzlement at the idea of "rainforests" burning (*see Box 4*), scientific evidence based on radiocarbon dating of charcoal deposits found in the soils of East Kalimantan indicates that forest fires have repeatedly burned areas of lowland rainforest, starting at least 17,500 years ago.[7] The earlier fires are believed to have been naturally caused during severe droughts, probably during the longer dry seasons that appear to have characterized Quaternary glacial periods.[8] Studies of buried pollen in Queensland, Australia, support the existence of alternating wet and dry climate phases in the past.[9] These studies indicate an almost complete absence of charcoal during more humid climate phases that correspond to the heights of the interglacial periods over the past 190,000 years. Climatologists believe that the Earth's climate is currently in a relatively wet phase that is not believed to be associated with major fires in tropical rainforests.

Humans probably had a role in starting forest fires in recent millennia and for tens of thousands of years may have deliberately burned forests to improve hunting. As prehistoric human settlers of the Indonesian archipelago began to switch from hunting and gathering to growing crops, they used fire to clear agricultural plots in the forest, a practice that has persisted until the present. (*See Box 5*.)

Forest fires on the islands of Borneo and Sumatra have been reported a number of times over the past 150 years. Large forest fires are reported to have occurred in what is now Central Kalimantan, on Borneo, in 1877, seriously affecting large areas of forest.[10] Grasslands still cover the 80,000-ha Sook Plains in Sabah (Malaysian Borneo) as the result of a drought-related forest fire in 1915.[11]

Periodic fires have been reported in the Danau Sentarum Wildlife Reserve in West Kalimantan since the middle of the last century.[12] The relatively fire-prone heath forests of Sabah and Sarawak (Malaysia) burned spontaneously or by human action in the 1880s, the early 1930s, 1958, 1983, and 1991.[13] The rainforests of Papua New Guinea are known to have burned during droughts, as indicated by the oral histories of indigenous peoples, charcoal buried in the soil, and historical accounts over the past century.[14]

The burning of Sumatra's and Kalimantan's forests is clearly not a recent or geographically unique phenomenon, but in the past neither naturally caused fires nor human use of fire led to significant deforestation; both islands remained largely forested until recent decades. Earlier fires were undoubtedly smaller in area and were probably more spread out over time than the fires of the past two decades. A 1924 forest map of what are now the provinces of Central, East, and South Kalimantan showed that 94 percent of this large portion of Borneo was still covered by forest. It is only

---

3 · EL NIÑO, DROUGHT, AND FOREST FIRES IN INDONESIA

El Niño is a periodic climatic phenomenon caused by interaction between the atmosphere and abnormally warm surface water in the eastern Pacific Ocean off the coast of South America. This sea temperature anomaly affects global climate, but its effects are particularly pronounced in Indonesia and other parts of the western Pacific, where droughts often result. El Niño events occur every 2 to 7 years, usually last about a year, and are sometimes followed by an unusually wet year.[1]

The severity of drought in Indonesia varies significantly from one El Niño to another. Particularly severe events result in major shortfalls in agricultural production,[2]

scarcity of surface water, and impacts on forests, including tree mortality and disrupted cycles of flowering and fruiting.[3] Over the past three decades the El Niño phenomenon has occurred in 1972, 1976,1982–83, 1987, 1991, 1994, and 1997–98. As an indication of the severity of the 1997–98 event, the Wanariset Forest Research Station in East Kalimantan received only 300 millimeters (mm) of rain in the 12 months ending April 1998, whereas annual precipitation averages 2,700 mm in that area.

Indonesian rainfall records dating back to the beginning of the 19th century reflect periodic droughts believed to have been caused by El Niño, but only recently

have scientists understood the mechanism that causes these events. Colonial records indicate that severe droughts affected agriculture and livestock production in the 19th century and resulted in forest fires.[4] Unlike the fires of the past two decades, earlier fires were usually caused naturally or were used for small-scale land clearing near what were then sparse forest settlements. During the 1997–98 drought, not only was the fire hazard very high as a result of extreme drought and heavy fuel loads in logged forests, but the risk that fires would spread out of control was higher because of the large areas of disturbed forest and scrubland close to extensive land-clearing operations. The El

Niño drought indeed increased the fire hazard, but human actions were the direct cause of uncontrolled fires. Poor forest management resulted in heavy fuel loads in logged or otherwise disturbed forest, and undisciplined use of fire for land clearance provided the flame that ignited the fuel.

Notes:

1. Nicholls, 1993.
2. Malingreau, 1987.
3. Wirawan, 1993.
4. Allen, Brookfield, and Byron, 1989.

Television viewers around the world were perplexed to see dramatic pictures of Indonesia's rainforests burning. These images clashed sharply with the lush green tropical rainforests featured in nature programs.

Rainforests burn because of a number of interdependent natural and human-related factors. These complex factors are often obscured by politically charged rhetoric, oversimplifications, and lack of factual information.

The danger that a forest will burn depends on the levels of fire hazard and fire risk, terms that are precisely defined by scientists who study forest fires.

■ Fire hazard is a measure of the amount, type, and dryness of potential fuel in the forest. Combustible fuel includes leaf litter,

low vegetation, grass, and dead wood in the form of logging wastes or fallen trees. The dryness of the fuel is related to how long vegetation has been dead or drought stressed, the period without rain, the relative humidity and temperature of the air, and wind speed.

■ Fire risk is a measure of the probability that the fuel will ignite. It is usually related to careless human actions, such as deliberate burning when fire hazard is high. Fire risk can be increased by natural factors such as lightning and by coal seams that catch fire. Abandoned logging roads provide easy access to otherwise remote forests, greatly increasing fire risk when settlers use fire for land clearance near forests.

Fire hazard can be rated with a reasonable degree of scientific accuracy.

Assessing the level of risk is much more subjective because human attitudes and motivations must be taken into account.

In the absence of drought, undisturbed mature rainforest is highly resistant to burning because of the high humidity below the forest canopy and the scarcity of fuel such as ground vegetation, leaf litter, and fallen branches.[1] Fires can start naturally in rainforests during periods of extreme drought; disturbed forest is much more fire prone. Forests adapted to growing on sandy and limestone-derived soils are more susceptible to fire than forests growing on other soil types. Peat-swamp forests are particularly vulnerable to above- and below-ground fires when water levels drop during droughts.[2]

Tropical rainforests recover even after a severe fire if they are left undisturbed and if there are seed sources nearby. Hundreds of years may be required to reach a successional stage approximating the species composition that existed prior to the fire. High-intensity fire followed by frequent burning leads to conversion of forest to grassland. The slow recovery of tropical rainforests after burning indicates that they are not well adapted to fire, unlike monsoon forest formations in seasonally dry parts of eastern Indonesia that recover quickly from frequent fires.

Notes:

1. Whitmore, 1984.
2. van Steenis, 1957.

---

since the advent of systematic logging and other forms of forest degradation in the late 1960s, coinciding with the establishment of the New Order regime, that fire has loomed as a large-scale and recurrent disaster.

THE GREAT KALIMANTAN FIRE OF 1982–83

The first collision between the periodic El Niño phenomenon and the Suharto regime's forest exploitation policies occurred in 1982–83 in the 210,000 square kilometer (km²) province of East Kalimantan. Starting in 1970, this vast province—almost completely covered by various types of rainforest, including dense stands of commercially valuable dipterocarp species—experienced an explosive timber boom. Nearly all of the province was divided up into large logging concessions, and by 1979 the annual cut had reached 9 million cubic meters (m³) of logs. Logging practices were wasteful and destructive, taking about 30 percent of basal area, damaging up to another 40 percent of forest stands, and leaving an enormous

accumulation of logging waste in the forest. Pioneer and secondary species sprouted rapidly in logged-over areas and on abandoned logging roads, forming a dense ground vegetation in place of the generally sparse ground cover found in primary rainforests.

A severe El Niño-induced drought struck the province between June 1982 and May 1983. At three sites where measurements were taken during this period, rainfall was only 30–35 percent of the normal amount. Lakes dried up, crops failed, river transport was cut off for many remote areas, and clean drinking water became scarce. By November 1982, most of the normally evergreen canopy trees had lost their leaves, and many trees had died. Temperatures were unusually high—an average 3°C hotter than normal at one monitoring station—further intensifying drought stress on the vegetation.[15]

In November-December 1982, fires started almost simultaneously across vast areas of the province. Since the forest was still relatively

moist at that point, these were relatively cool ground fires, creeping slowly along the forest floor and not causing a great deal of damage. After a brief respite from drought at the end of December, the situation rapidly deteriorated. The first wave of fires had amplified the effects of the drought, drying out the ground vegetation and the understory of the forest and increasing the amount of litter. Accumulated logging waste added to the easily combustible layer of material that covered much of the forest floor. When the fires began again, much of the province became an inferno. Canopy trees burned like torches, and whole trees exploded when their resin vaporized and was ignited. By April 1983, aerial transport in the province had come to a complete halt, and the sun was blotted out in a perpetual shroud of smoky haze.[16]

By the time rains finally came in May 1983, 3.2 million ha—an area the size of Belgium or Taiwan—had burned; of this, 2.7 million ha was tropical rainforest.

Damage from the fires varied in different areas, from creeping ground fires in primary forests to complete destruction of newly logged areas and peat-swamp forests. Some 730,000 ha of the commercially valuable lowland dipterocarp forests were badly damaged, and another 2.1 million ha were lightly or moderately damaged.

Droughts and fire have been a feature of East Kalimantan's landscape for millennia. What caused these fires to metastasize into what was, at the time, the largest forest fire ever recorded? A comprehensive field study of the fires carried out in 1983–89 with the support of the German Agency for Technical Cooperation, pointed out that "it was not the drought which caused this huge fire, it was the changed condition of the forest" due to widespread and reckless logging activities over the previous decade.[17] Logging transformed the fire-resistant primary rainforest into a degraded and fire-prone ecosystem. The drought then set the stage for catastrophe as

SHIFTING CULTIVATION AND FIRE IN INDONESIA

The agricultural system based on a cycle of forest clearing, cultivation, and fallowing, called swiddening or shifting cultivation, has been adopted throughout most of the Indonesian archipelago over a period of thousands of years.[1] Swidden cultivation has continued into this century in parts of Indonesia where soils are too poor to support permanent cultivation of annual crops. Until recently, swiddening was the dominant form of cultivation in Kalimantan, and it is still practiced there, as well as in Sumatra and other areas of Indonesia outside the densely populated islands of Java and Bali.

A major labor effort is required to clear mature forest, so swidden plots are usually limited to less than 1 ha when hand tools are used for clearance. In recent years, chain saws have made it possible for one family to clear significantly larger plots. Typically, swidden plots are cultivated for one to three years. They are then left fallow for several years to allow natural vegetation to regrow, creating a mosaic of pioneer and secondary vegetation patches in the mature forest.

Suppression of swidden cultivation and its replacement by irrigated rice was a major feature of Suharto-era forest and agricultural policy. Traditional swidden systems are well adapted to poor soils, low land-to-labor ratios, and the livelihood needs of rural communities in the areas outside Java and Bali. These systems, however, contradict the irrigated rice-based system of the dominant Javanese culture, and they presented an obstacle to the exploitation of forest lands and resources by outsiders that was promoted by the regime. Echoing colonial assessments and Javanese cultural biases, the government has long maintained that swidden cultivation and its practitioners are environmentally destructive, backward, and wasteful and has often blamed swiddeners for outbreaks of fire in forest areas.[2] As traditional swidden systems have eroded or become more intensive, shifting cultivation has become a much more negative environmental factor. In areas with growing populations of forest dwellers, the number of years that a swidden plot lies fallow has been shortened so much that regeneration does not progress beyond pioneer vegetation. This trend has been accelerated by the desire to grow cash crops. Traditional swidden farming at low population densities has only a slight impact on forest biodiversity compared with the accelerated system currently practiced in many places. Extreme shortening of swidden cycles can result in the conversion of forest to grassland, which may then be burned annually to maintain grassland for improved cattle grazing or to facilitate hunting.

Notes:

1. Marten, 1986.
2. Dove, 1985.

"small agricultural fires . . . escaped their bounds into nearby dry secondary and logged-over forests."[18]

Burning for land preparation is practiced as part of many agricultural activities in the area, including cash-crop farming, subsistence upland rice cultivation, and preparation of garden plots.[19] (*See Box 5.*) Although some of the fires were undoubtedly started by traditional subsistence farmers, many were set by peppercorn and other cash-crop growers and by land speculators. Between 1970 and 1980, East Kalimantan's population had doubled as a result of the timber and oil booms, and many spontaneous migrants had arrived in the province to stake out cash-crop plots on the forest frontier, often following logging roads into the interior.[20] This influx was intensified by the beginnings of the government's transmigration program, which brought about 91,000 new settlers from Java and Bali to the province between 1970 and 1983.[21]

Evidence from the German-assisted survey demonstrates conclusively that logging was the primary reason for the extent and severity of the 1982–83 fires. Only 11 percent of undisturbed primary forests in the areas affected by the drought and fires actually burned. Even there, only ground vegetation burned, and the forest had completely recovered by 1988. By contrast, in an area of nearly 1 million ha of "moderately disturbed" forest (80 percent of which had been logged prior to the fires), 84 percent of the forest burned, and the damage was much more severe. According to the study, "The standing stock is heavily reduced, future exploitation of these stands, with the exception of the dead hardwoods, will not be possible within 70 to 100 years and even then the rate of exploitation will be far below the exploitation rate of undisturbed forests of today."[22] In an area covering 727,000 ha of heavily disturbed forests (70 percent of which had been logged within 8 years before the fires), 88 percent burned, and fire completely destroyed the forest structure, meaning that "natural succession will need several hundred years to reach the stage of a typical tropical rainforest ecosystem."[23] The researchers concluded that "it is obvious that logging shortly before the fire had the most influence on the degree of damage."[24]

Widespread fires reoccurred a number of times in the decade following the great Kalimantan fire, burning an estimated 500,000 ha in 1991 and nearly 4.9 million ha in 1994.[25] Haze from the fires blanketed Singapore and Malaysia, as well as large areas of Indonesia, disrupting air and sea transportation. In the aftermath of both fire episodes, the government adjusted its policies and established a variety of new fire-control programs and committees, at least partly in response to concerns voiced by neighboring countries. A number of international aid agencies provided support for fire-related programs.[26] In 1997, however, it became painfully evident that while these efforts had boosted capacities to predict and monitor fires, they had done virtually nothing to strengthen Indonesia's ability to prevent or combat fires. Even more important, the two basic causes of recurrent fires—changes in vegetative cover caused by deforestation processes and the practice of using fire to clear land—had not been dealt with at all.

By 1997, the processes of deforestation and land degradation unleashed by the Suharto regime had intensified and diversified into the clearing of vast areas for timber and oil palm plantations, in addition to the continuing destruction wrought by logging operations. When the fiercest El Niño-related drought in at least a century swept across the archipelago in mid-1997, it heralded a conflagration that would dwarf the East Kalimantan disaster and dismay the world.

By early 1997, oceanographers and atmospheric scientists were predicting that 1997 would be an El Niño year. In June the trade winds reversed their direction across the Pacific, and by early July sea surface temperatures off the coast of South America were already 4°C above normal—clear signs that an El Niño was starting.[27] These phenomena coincided with the normal dry season in Indonesia. Despite warnings by the environment minister, burning continued across vast areas of Sumatra and Kalimantan to clear vegetation in preparation for planting crops and trees, a practice that had escalated dramatically in recent years. The first fires were picked up on satellite images in January 1997 in Sumatra's Riau province, and the fires increased in number and distribution as the dry season began.

The use of fire for land clearance is not restricted to Borneo and Sumatra—fires were reported from 23 of Indonesia's 27 provinces in 1997-98—but by July the large number of fires set on those two massive islands by plantation firms and government projects clearing tens of thousands of hectares at a time had produced enough smoke to create a blanket of haze that spread hundreds of kilometers in all directions. Deliberately set fires in grasslands and scrublands escaped into adjacent logged forests that burned with greater intensity. The fires eventually reached drained peat swamps, where fires burned beneath the surface long after above-ground fires had exhausted their fuel supplies.

Fires on other islands such as Java and Sulawesi were smaller and had more localized impacts. Irian Jaya, a vast Indonesian province occupying the western half of the island of New Guinea, was badly affected by the drought, which caused hundreds of deaths from waterborne diseases, malaria, and food shortages. Fires also burned there, but the total area affected was much smaller than in Kalimantan and Sumatra. Haze from the Irian fires, however, periodically spread as far as Darwin in northern Australia. (*See Map 2.*)

Large-scale burning has produced persistent haze over large areas of Sumatra and Kalimantan during every dry season, but the haze normally dissipates in September, when heavy rains extinguish the fires. This was not the case in 1997. The rains failed, the fires intensified, and the haze thickened and spread to neighboring countries. Haze reached Malaysia and Singapore in July, and air quality deteriorated dramatically in September, triggering an outburst of complaints that drew global media attention.

By late September approximately 1 million km² were haze covered, affecting about 70 million people. Land, air, and sea transport accidents, including a ship collision in the Straits of Malacca that killed 29 people, were linked to the poor visibility caused by the haze. Hospitals and clinics were filled with people seeking treatment for a variety of respiratory, eye, and skin ailments. Schools, businesses, and airports closed, and tourists stayed away, inflicting severe economic hardship on the region.

THE GOVERNMENT RESPONSE

Even as fires burned out of control into surrounding forests, peat swamps, and agricultural lands, plantation owners and farmers started new fires to take advantage of the extremely dry conditions. This caused the haze to intensify and spread further, resulting in health alerts and transportation disruptions across the region. The government announced a total ban on burning in mid-September, followed by threats to punish offending firms. President Suharto publicly apologized on two occasions to neighboring countries for the haze and demanded that Indonesians stop illegal burning. These apologies were particularly embarrassing because in 1995 Suharto had assured Malaysia and Singapore that transboundary air pollution such as had occurred during the 1994 El Niño drought would not be repeated.

In late September the minister of forestry released the names of 176 plantation, timber, and transmigration land-clearing firms suspected of deliberate large-scale burning within their work areas. The suspect firms were identified by comparing hot spots identified by U.S. National Oceanic and Atmospheric Administration (NOAA) satellites with Ministry of Forestry maps of timber and plantation concession areas. The firms were given two weeks to prove that they were not guilty of illegal burning or risk revocation of their timber-cutting licenses (essentially, a land-clearing license in this context). A number of licenses were revoked but were mostly reinstated in December. As of mid-May 1998, not a single company or person ordered by a company to clear land by burning had been brought to trial.[28]

> *During the fires, hospitals and clinics were filled with people seeking treatment for a variety of respiratory, eye, and skin ailments.*

In an important test case in October 1998, PT Torus Ganda, a firm with operations in Riau province, was taken to court by the Ministry of Forestry for destruction of the forest by burning. Expert testimony by the Environmental Management Bureau (BAPEDAL) of the Ministry of Environment was reportedly not taken seriously by the court, which exonerated the plantation owners on all charges. The firm's operations were then frozen by decree of the Riau governor pending action by the company to rectify its land-clearing practices, but in July 1999 the local press reported that the company was ignoring the decree and conducting business as usual.[29]

In another test case, in 1998 the Indonesian Forum for the Environment (WALHI), a coalition of Indonesian nongovernmental organizations (NGOs), brought a civil suit under the new 1997 Environmental Law against 11 firms alleged to have illegally burned to clear land in southern Sumatra. WALHI sought damages of Rp 11 trillion (more than $1 billion) to be paid to the state to rehabilitate burned areas. Detailed geographic information systems (GIS) information was presented but was thrown out by the court, leaving only eyewitness testimony. Two of the firms were found guilty, but the court merely directed them to pay court costs, correct their fire management, and establish a fire brigade.[30]

The message of these cases is clear, as noted by the study on the fires by Indonesia's National Development Planning Agency (BAPPENAS) and the Asian Development Bank (ADB): "These cases could well have far-reaching consequences and may seriously undermine other attempts at further prosecutions. Even more significantly, the first case calls into question the capacity of the government to issue instructions with sanctions against environmental damage if these instructions cannot be properly enforced in the courts."[31]

Sporadic firefighting efforts by the Indonesian government with assistance from Malaysian volunteers and fire suppression aircraft from Australia and the United States were largely ineffective. Poor coordination (especially between air and ground operations), lack of equipment, lack of funds, insufficient training, lack of water, and the remote location of many of the fires were often cited as the reasons for failure. Aerial suppression by water bombers was hindered by the lack of accurate land cover maps and infrastructural support, and land-based efforts were impeded by the reluctance of many rural people to fight fires on land that was not theirs.[32] The number of fires began to decline during October and November, probably partly due to mounting pressure exerted by the government on plantation firms but also because the firms had burned as much land as they needed by that time. Peat swamps were still burning in late November, but these fires were partially extinguished when rain finally began to fall in December.

In December, the Association of Southeast Asian Nations (ASEAN) adopted a Regional Haze Action Plan under which Indonesia pledged to improve its firefighting capability. [33]

The rainy season, which usually lasts at least six months in western Indonesia, began to taper off in less than two months. By mid-January 1998, new fire hot spots began to appear on NOAA weather satellite images as the drought carried over into a second calendar year and a new rainfall cycle. The pattern of 1997 was repeated in the coastal swamps on Sumatra's east coast from January through April. In Kalimantan the fires were concentrated in East Kalimantan, a province that had not been extensively burned in 1997. The drought was also beginning to cause food shortages due to below-normal harvests and total failure of the rice crop in some areas. The plight of rural communities, already reeling from the effects of the fires, haze, and drought, was worsened by the growing economic impact of the dramatic devaluation of the Indonesian currency over the second half of 1997.[34] Farmers began to clear even more land by burning, in the hope that they could increase the next harvest to make up for 1997 losses. Fears arose that forest exploitation and related burning would increase as firms tried to offset the effects of the economic crisis. [35]

By the end of January 1998, hundreds of hot spots, concentrated in coastal areas of East Kalimantan and the coastal peat swamps of Riau and North Sumatra provinces, were evident on satellite images. These hot spots indicated the locations of newly set fires, smoldering peat that had burst into flame, and continued underground burning of coal seams. The most extensive burning in January took place in East Kalimantan's Kutai National Park, already badly damaged by previous fires, logging, mining, and agricultural encroachment.

By mid-February the fires were headline news internationally again as haze returned to parts of Borneo and Sumatra, resulting in respiratory problems and domestic airport closures.[36] Neighboring countries began to fear a return of the haze that had blanketed the region only three months earlier and were not reassured by the Indonesian government's weak efforts to prevent or extinguish new fires. The army had agreed to take a more active role in fighting the fires than it had in 1997, but it later reduced the number of troops assigned to this duty in order to prepare to combat civil unrest resulting from the economic crisis.[37]

Fires continued to spread during March, and at the end of the month a Japanese remote-sensing system indicated that there were as many as 5,000 hot spots on the island of Borneo, while other sources reported over 1,000. The Southeast Asian environment ministers met in Brunei during the first week in April to discuss the fires, the third such meeting in four months. The ministers concluded that firefighting efforts should be focused on containing existing fires and preventing new outbreaks.

Efforts to fight fires were hampered by increasing water scarcity because the drought caused surface water to dry up and the groundwater level to sink below the reach of wells. In mid-April, a United Nations Disaster Assessment and Coordination team estimated that an effective firefighting effort would require at least 10,000 firefighters supported by water bombers, but that sustained rain provided the only hope of extinguishing the fires completely. The difficulty of firefighting under such extreme drought conditions is illustrated by the experience of the staff of the Wanariset Forest Research Station in East Kalimantan in fighting repeated fires in the station's 3,500-ha research forest. Despite their vigilance, only 20 ha remained unburned by mid-April.[38]

The newly installed environment minister, Juwono Sudarsono, estimated that effectively fighting the East Kalimantan fires would cost $2 billion.[39] The minister compared the lack of government control over the fires in East Kalimantan to the lawlessness of the American Wild West in the 19th century and regretted his inability to create a sense of urgency among government officials.[40] A week later he admitted that the forest fires were a low priority for the Indonesian government, which was more concerned with countering the effects of the economic crisis.[41]

Beginning in March 1998, Sarawak and Brunei were affected by the haze, and Malaysia and Singapore worried that the normal shift of the monsoon winds from west to east in May would again blanket the Malay Peninsula in smog. Brunei, which had escaped the haze in 1997, took strong health precautions, including closing schools for two weeks. By mid-April, authorities in Sarawak were again considering declaring a state of emergency because of the soaring air pollution levels, and schools were closed in several towns in the state.[42] The World Meteorological Organization reported at the end of March that El Niño conditions would abate sometime during June–August, but since those months correspond to the middle of the Indonesian dry season, some meteorologists feared that rains heavy enough to extinguish the fires might not fall until October.

Heavy rains did fall during the first part of May, extinguishing many of the fires in Kalimantan and Sumatra, but drought conditions returned toward the end of the month. By late May, consensus was growing among oceanographers and atmospheric scientists that El Niño conditions were abating and there was a shift toward ocean conditions that usually precede La Niña, a climatic phenomenon that usually causes above-average rainfall in Indonesia for a year or more.[43] Heavy rains began in June and led to floods in East Kalimantan in July.

In late 1998, hot spots again began to appear on NOAA and other satellite data for Sumatra and East Kalimantan, indicating that fire-setting behavior had not changed much despite the recent experience.[44] Meanwhile, haze briefly reappeared in Singapore, Sarawak, and southern Peninsular Malaysia in late November.[45]

It is difficult to determine precisely the total area burned during the 1997–98 fires or to estimate what vegetation types burned in which areas. (*See foldout map on inside cover.*) On the basis of the most recent estimates available in early 1999, it seems certain that at least 9.7 million ha burned. (*See Box 6 and Table 1.*)

The extent of the area affected by air pollution from the fires is easier to determine. Indeed, the international news media were initially attracted to the 1997 fires by the dramatic spectacle of a "thousand-mile shroud" spreading over an area of 1 million square kilometers inhabited by hundreds of millions of people. Ramon and Wall (1998: 3) observed that "whereas the impact of fires concerns mainly foresters and conservationists, it is the smoke that causes politicians and economists to react." In one Indonesian town, it was reported, schoolchildren were tied to a rope to prevent them from becoming lost in the haze on their way to school. In April 1998, the 1,788-room palace of the Sultan of Brunei was reported to have been almost invisible behind a thick curtain of smog,[46] as was Kuala Lumpur's landmark Petronas Twin Towers, the tallest building in the world. Haze-related transportation accidents were widely reported, as well.

Mixtures of visible suspended airborne chemicals normally associated with urban air pollution are called smog, but government officials in the region were anxious to downplay the connection by calling it "haze." Malaysian information minister Mohammed Rahmat went a step further in April 1998 by

warning his nation's broadcast media not to use the word haze either, or risk having their operating licenses revoked.[47] In August 1999, the Malaysian government went even further, making information on air quality an "official secret" and directing the firm awarded the concession to monitor air pollution to make its readings available only for "private consumption." The environment minister stated that the measure was taken so as "not to drive away tourists."[48]

Indonesia does not routinely monitor air pollution levels, but Malaysia and Singapore do. A reading of 100 on the standard air pollution index (API) is considered unhealthy; 300 is hazardous. API readings remained in the hazardous range for long periods in September and October 1997 in the Malaysian state of Sarawak, with a high of 849 recorded. A reading of 1,000 was recorded in the interior of East Kalimantan in early October 1997, a level that was probably not unusual in areas close to the fires.[49] Malaysians and Singaporeans were informed when air pollution reached unsafe levels and were warned to take appropriate protective measures, but most Indonesians were unaware of the level of health hazard.

TABLE 1

Estimated Extent of Spatial Damage by Fire, 1997–98 (hectares)

ISLAND	MONTANE FOREST	LOWLAND FOREST	PEAT & SWAMP FOREST	DRY SCRUB & GRASS	TIMBER PLANTATION	AGRICULTURE	ESTATE CROPS	TOTAL
Kalimantan	0	2,375,000	750,000	375,000	116,000	2,829,000	55,000	6,500,000
Sumatra	0	383,000	308,000	263,000	72,000	669,000	60,000	1,756,000
Java	0	25,000	0	25,000	0	50,000	0	100,000
Sulawesi	0	200,000	0	0	0	199,000	1,000	400,000
Irian Jaya	100,000	300,000	400,000	100,000	0	97,000	3,000	1,000,000
Total	100,000	3,100,000	1,450,000	700,000	188,000	3,843,000	119,000	9,756,000

Source: BAPPENAS, 1999.

WHAT BURNED AND WHERE?

Obtaining accurate data about the spatial distribution of the 1997–98 fires, the total area burned, and the proportion of different vegetation or land-use types that burned is difficult because of the size and wide distribution of the burns, the remoteness of many of the sites, the inability of most satellite remote-sensing devices to penetrate the thick haze while the fires are burning, and the need to verify interpretations of images.

The 1982–83 fires in East Kalimantan, for example, were not detected by remote-sensing satellites for almost three months.[1] Basic descriptive information was not compiled until 1984, when a relatively rapid aerial and ground survey of the affected area was completed.[2] The findings were revised after a later and more thorough analysis in 1989.[3] The 1997–98 fires occurred over a much wider area of the country, and compiling authoritative data on exactly what burned, where, and how badly will take years.[4]

During the 1997–98 fires, the Ministry of Environment and several other organizations tracked where fires were burning each day by monitoring hot spots that appeared on NOAA weather satellite images. These data are received directly from the satellites at several stations in Indonesia. The NOAA data (also used to monitor the 1982–83 fires in East Kalimantan) can be used to assess hazard by revealing the dryness of vegetation and ground temperature patterns.[5] Hot spots that appear on NOAA satellite images provide a general picture of the distribution of fires on a given day but indicate little about the size of the area burned and nothing about what burned.

Remote-sensing experts working on a European Union–funded Forest Fire Prevention and Control Project (FFPCP) made a preliminary estimate, using satellite imagery in sample areas, that the 1997 fires burned 2.3 million ha in South Sumatra province alone.[6] The fires were almost evenly divided between wildfires and controlled burns. The project found that the types of vegetation that burned, in descending order of importance, were wetland vegetation being cleared to prepare rice fields, secondary brush, scrublands and herbaceous swamplands, dryland shifting agriculture plots, and grassland in coastal peat swamps.

The Singapore Centre for Remote Imaging, Sensing, and Processing (CRISP), using SPOT (Système pour l'observation de la Terre) satellite imagery, calculated that in 1997 approximately 1.5 million ha had burned in Sumatra and approximately 3.0 million ha in Kalimantan.[7] CRISP concluded that most burning occurred in lowland areas near rivers and roads; montane forests were virtually untouched by fire in 1997.

Another analysis, carried out by experts collaborating with WWF-Indonesia calculated that between 1.97 million and 2.3 million ha burned in Kalimantan during August-December 1997.[8] The Ministry of Forestry and Estate Crops, however, officially estimated that in the country as a whole, only 165,000 ha of designated forestlands had burned in 1997.[9] Forestry officials, however, kept track of fires on the basis of unverified reports from timber concessionaires and plantation owners, who have no incentive to report fires accurately.

The WWF-Indonesia study found that the 1997 Kalimantan hot spots were most frequently found in peat swamps and other wetlands but that lowland forests had the highest number of detected fires.[10] Using GIS to correlate hot spots with human and natural features, they found that fires tended to be clustered near rivers and agricultural lands, not necessarily close to settlements—supporting the hypothesis that the fires were set to clear land for commercial plantations. Ground-based observations and interviews with local people indicate that smallholder plantations and home gardens, as well as oil palm and other commercial plantations, were consumed by the fires.[11]

In late 1998, CRISP announced that its analysis of satellite data of the 1998 fires revealed that 2.5 million ha had burned in East Kalimantan and 500,000 ha in Sabah, Malaysia. Combined with CRISP's estimates of area burned in 1997 in Kalimantan and Sumatra, the total for both fire episodes approaches 8 million ha.[12] The East Kalimantan-based Integrated Forest Fires Management Project estimated that between 4 million and 5 million ha had burned in East Kalimantan alone, mostly in 1998.[13] This figure was revised upwards in 1999 to 5.2 million ha, based on the results of a detailed mapping exercise.[14] (*See Box 7.*) The discrepancies between the findings of the two groups reflect the technical constraints inherent in this type of remote-sensing analysis. The important points are that very large areas of Sumatra and Kalimantan burned in 1997 and 1998 and that many different types of vegetation burned.

In the first part of 1999, a technical team funded by the ADB and working through BAPPENAS aggregated and analyzed all available data sources and estimated that the area burned during 1997–98 totaled more than 9.7 million ha, as noted in Table 1.

Notes:

1. Malingreau, Stephens, and Fellows, 1985.
2. Lennertz and Panzer, 1984.
3. Schindler, Thoma, and Panzer, 1989.
4. For a discussion of the difficulties inherent in accurately determining areas burned, see Fuller and Fulk, 1998.
5. Malingreau, Stephens, and Fellows, 1985.
6. Ramon and Wall, 1998.
7. Liew and others, 1998.
8. Fuller and Fulk, 1998. The discrepancy between the CRISP and Fuller and Fulk estimates may be due to differences in data, methods, and coverage between the SPOT and NOAA satellites.
9. GOI, Ministry of Forestry and Estate Crops, 1998.
10. Fuller and Fulk, 1998.
11. Gonner, 1998; Vayda, 1998; Potter and Lee, 1998a.
12. *Straits Times*, November 23, 1998.
13. Schindler, 1998.
14. Statement of Lothar Zimmer, German Federal Ministry for Economic Cooperation and Development, Consultative Group on Indonesia Meeting, Paris, July 28-29, 1999.

THE POLITICS OF FIRE

As soon as the Indonesian forest fires and the resulting haze became headline news in September 1997, a fierce political battle began about who was to blame for the disaster and what the most effective response would be.

The Indonesian government and its allies in the forestry and agribusiness industries have traditionally blamed forest fires on small-scale shifting cultivators and on periodic droughts and other vagaries of nature. In the aftermath of the 1982–83 fires in East Kalimantan, despite ample evidence that poor logging practices were the main factor in creating the conditions for the huge conflagration, Forestry Minister Sudjarwo told the press that "nomadic cultivators who use the outdated [slash and burn] techniques" were the primary cause of the fire and that logging companies were "not responsible." Queried on the role of the vast quantities of waste and dead trees left in the wake of logging operations, he countered that "to be fair to them, it is not economical for them to clear the forest [after logging]. The cost of clearing is very high." He went on to suggest that the fire might be a blessing in disguise, since the government wanted to clear a great deal of land in the province, and with the fires "what you have is land clearing for free."[50]

By 1997, the attitude of many government officials had become less defiant, but Sudjarwo's line was still being echoed by Bob Hasan, an outspoken timber baron and crony of President Suharto. In October, Hasan told the press that the extent of the fires had been exaggerated and that logging concessions bore no responsibility for the burning.[51] Later that month he said that activists calling for sanctions on companies found to be using fire to clear land were influenced by communist agents.[52]

The power of Hasan and other entrenched timber industry cronies largely blunted the good-faith efforts of Forestry Minister Djamaluddin Suryohadikusumo to take action against companies accused of setting fires, and Djamaluddin was widely rumored to be close to resigning over this predicament. (Suharto appointed Bob Hasan minister of forestry soon thereafter, in the short-lived cabinet that took office in March 1998.)

The forest fires of 1997 opened the way for a number of politically marginalized actors to take on important roles in influencing public perceptions and opinions about the fires and about Indonesia's natural resource management policies in general. For a brief period (August-October 1997), Indonesia's mass media were dominated by marginal actors, including the minister of state for environment, the Environmental Impact Assessment Agency (BAPPEDAL), the Meteorological and Geophysical Agency, the national space agency (LAPAN), and a variety of NGOs. Statements coming from timber and plantation firms, the Ministry of Forestry, and the Directorate General for Plantations in the Ministry of Agriculture took on a defensive and reactive character. The Transmigration Ministry and the armed forces chose to keep silent.

The most outspoken and consistent opinion leader in this period was the minister of state for environment, Sarwono Kusumaatmadja. Beginning in early September, and taking advantage of the pressure from Malaysia and Singapore to do something about the haze, Sarwono was able to push the issue of the forest fires to the top of the public and media agenda. As a result, Indonesia was obliged, for the first time, to take the fires seriously. On September 8, President Suharto officially ordered that use of fire to clear land be stopped, and on September 16 he publicly apologized to neighboring nations at the opening of an ASEAN environment ministers' conference. These actions put government agencies on notice that they had to act.

Building on this momentum, Sarwono took the lead in fingering oil palm and timber plantation companies as the major culprits in creating the disaster, using overlays of NOAA hot spot maps with maps that showed the location of the plantation concessions. At the outset he encountered stiff resistance from government forestry and plantation agencies. But public and media pressure grew, and in mid-September the Ministry of Forestry announced that 176 oil palm plantations, industrial timber estates, and transmigration sites were suspected of intentionally and systematically using fire to clear land in Sumatra and Kalimantan.[53] In early October, the ministry canceled 166 timber-cutting rights held by these firms, though it did not suspend or cancel any actual concession agreements.

This action gave rise to considerable resistance from the industries, and 121 firms sent rebuttals to the ministry.[54] But the efforts of the timber and plantation industries to influence public opinion tended to be defensive because the public had in large part already judged them to be the guilty parties. Like some government officials, many firms continued to accuse shifting cultivators of setting the fires. But, in October, several of the biggest timber barons—and also members of the Indonesian Business Council for Sustainable Development—announced that they would contribute 20 billion rupiah (about $5.5 million at the exchange rate at that time) toward efforts to fight the fires and provide assistance to affected communities. Details on the actual disbursement and use of these funds were never made public, however.

Apart from the two environment agencies, a number of other previously marginal government agencies assumed important roles in shaping public opinion during this period. While ultimately unsuccessful, the cloud-seeding projects carried out by the Technology Development Agency (BPPT) and the armed forces across parts of Sumatra and Kalimantan that began in September served as a public symbol that the government was actually doing something about the fires. The activities of BMG and LAPAN in providing remote-sensing and other data on the drought and the fires gave those two agencies an unaccustomed public role. Indeed, it was their work, in cooperation with the environment agencies, that provided convincing evidence that plantation owners were major culprits behind the fires.

NGOs in Jakarta played an unprecedented role in influencing public opinion and policy through the mass media. WALHI, for example, produced a widely reported stream of press releases rebutting government and industry statements concerning who was setting the fires, their extent, and the probable economic losses. The fires gave WALHI and other NGOs a useful platform for their long-standing critique of the government's industry-oriented and destructive forest policies.[55] While WALHI and other Jakarta-based groups played an important role in advocacy, many local NGOs and youth groups undertook the distribution of face masks and medicine in the country-side at a time when government relief efforts were largely paralyzed, and many groups physically fought fires in the field. In addition, many NGO staff volunteered to help the Environment Ministry operate forest fire "command posts" to monitor reports from the field, analyze satellite data, and coordinate distribution of assistance.

The role of the press in shaping public opinion and influencing policymakers about the fires was extremely important and probably represented the Indonesian news media's most outspoken and influential performance to date, at least since Suharto came to power. Press coverage reached a peak of intensity from mid-September to mid-October, with events such as Suharto's September 16 apology to ASEAN, release of the list of companies suspected of intentionally setting fires, the arrival of Malaysian firefighters in Sumatra, declaration of a state of emergency in Sarawak, the crash of an Indonesian airliner in Sumatra, and the outbreak of numerous fires on the island of Java.

NGOs in Jakarta played an unprecedented role in influencing public opinion and policy through the mass media.

By the middle of October, however, press coverage fell off, although the fires continued to burn until the end of November. One journalist noted that "the fires just weren't news anymore," and at the same time the government decided to place the blame for the fires squarely on El Niño. During the first week of October, senior figures from the print and electronic media were summoned to a meeting with a number of key ministers held at the Ministry of Information.[56] The editors were told that henceforth they were to blame the forest fires on El Niño and cease the "polemics" that had characterized news coverage of the disaster up to that point. Only the Ministry of Environment and its Environmental Impact Agency resisted this line, telling the press in mid-November that "the 1997 forest and land fires are largely due to land clearance activities using this method [fire]. El Niño is only an additional factor. El Niño does not start fires, but only makes forests susceptible to fire."[57]

The evolution of public opinion about the forest fires had three distinct phases. First came realization that there was a serious problem and that it was affecting neighboring countries. In the second phase the press and public blamed the timber and plantation companies (and, by extension, the government agencies responsible for regulating them), and the industries concerned mounted strenuous efforts at rebuttal. In the final phase, faced with its own apparent inability to respond effectively to the fires, the government made a vigorous effort to convince the public that the fires were the result of a natural phenomenon occurring on a global scale and thus were legitimately outside the scope of human intervention.

Preventing and fighting fires on forestlands outside of densely populated Java has never been a very high priority for the government, and its regulatory approach has been reactive rather than preventive. Despite provisions in the 1967 Basic Forestry Law calling for development of regulations to deal with forest fires, the government only began this effort in the mid-1980s, in the aftermath of the great East Kalimantan fire of 1982–83. Additional regulations were issued following renewed large outbreaks of fire in the early 1990s. As late as April 1997, however, the ministry issued a regulation specifying procedures for "controlled burning" to clear land and only voided it in another flurry of reactive decrees when haze from the fires enveloped the region in September 1997.

The on-the-ground government response was also weak and equivocal. By October 1997, the government had essentially given up trying to put out the fires and was waiting for the rains to return. Assistance efforts offered by a number of nations (planes from the United States, firefighters from Malaysia) were dogged by poor support and coordination on the Indonesian side. Thousands of face masks donated by the United Nations Children's Fund (UNICEF) in October sat in the National Disaster Relief Office in Jakarta for weeks due to bureaucratic snafus in the distribution. An October 14, 1997, editorial in the *Bangkok Post* entitled "Indonesia's Shame Won't Blow Away" summed up the impatience and anger in the region at the government's ineffectual and defensive performance:

"For weeks, Indonesian big business cleared forests for palm oil cultivation, creating pollution that enshrouded neighboring states, causing inestimable damage to the health of millions, devastating agriculture and local economies alike. After trying to pin the blame on El Niño, an apology of sorts came but without a pledge to stop the seasonal devastation of the forests. And so the region can expect varying degrees of airborne delayed death next year too, and the year after that, and the year after that. It is abundantly clear that the region will be laid to waste as long as governments remain enslaved to big business."

Meanwhile, international aid agencies fell over each other trying to respond to the disaster, with each, it appeared, attempting to establish itself as the agency taking the leadership role in the crisis. One senior government official complained privately that many of the bilateral offers of aid were just "attempts to get Indonesia to buy expensive firefighting equipment from firms in their country." A July 1998 review listed more than a dozen projects by as many agencies, all claiming to be responding to the fires.[58]

Viewed through a political lens, the forest fires presented an unprecedented opportunity for hitherto marginalized actors to publicly raise fundamental questions about the destructive relationship between government natural resource policies and the crony cartels that dominate natural resource exploitation in

Indonesia. This window of opportunity for the critics did not last long, however, and basically slammed shut when the rain began to fall. The fires vanished from the press, and the effects of the fires on the health of the millions of people in Sumatra and Kalimantan who for months lived under a blanket of smog largely

vanished from policymakers' radar screens. The fall of Suharto in mid-1998 and the advent of the *reformasi* movement have once again opened that window of opportunity. The extent to which this opening is likely to be used to carry out meaningful forest policy reforms is discussed in Part II.

7

THE POLITICS OF DATA MANIPULATION: HOW MUCH OF EAST KALIMANTAN BURNED DURING 1997–98?

The official position of the government, as of August 1999, was that some 520,000 ha had burned in East Kalimantan province, and this figure was used by then-President Habibie in a speech in East Kalimantan that month. Unfortunately, extensive remote sensing work, confirmed by ground-checks carried out by the German-supported Integrated Forest Fires Management Project based in East Kalimantan, produced convincing and comprehensive data in mid-1999 showing that the fires had in fact covered some 5.2 million ha, ten times the government's official figure.[1]

Informed of this considerable discrepancy, the Minister of Forestry and Estate Crops, in early July 1999, wrote to the governor of East Kalimantan with respect to this "data gap," and in essence instructed him not to allow the new data to be made public: "Remembering current political developments, we think it is necessary to protect national stability. Let's not cause further debate [by making the data public] that could harm preparations for the

upcoming session of the People's Consultative Assembly [MPR]."[2]

In a meeting with staff of the German-funded project several days later, the governor indicated that he found the results of the new study credible, but requested that they not be made public in order to maintain "peace in his province." The project representatives reported that "the Governor refused to accept that a pro-active role of the Government, by accepting and using the results to implement the further necessary steps, would be beneficial to all parties" and concluded that "the letter of the Minister as well as the position of the Governor are difficult to comprehend and could cause serious problems for the future implementation of both [German-assisted fire-related] projects."[3]

This concern was reiterated by Germany's representative to the 1999 Consultative Group on Indonesia—the annual meeting of Indonesia's international donors—in Paris several weeks later, where he stated:

"All in all, the forestry sector is the most important of the focal areas of our cooperation. . . . It would therefore be highly irritating if recent reports from Indonesia were even just partially accurate: . . . .We are puzzled as to why the extent of recent fire-inflicted damage to the forest differs by a factor of 10, i.e. official estimates refuse to acknowledge the full extent of this catastrophe, namely that the burning of 5 million hectares in 1997-98 produced nearly one third of the world's total CO_2. . . .The EU Commission has prepared a draft resolution on this matter and I strongly appeal to you to give it the highest attention. Failing this, my government would be obliged to reconsider the future of our ongoing forestry projects. . . . I need not elaborate on the ecological but also international implications of such a reversal in the— so far—reform-oriented forest policy."[4]

Despite the exposure of this cover-up and the serious concerns publicly expressed by a senior official of one of Indonesia's major aid donors, the government had not,

as of October 1999, made the accurate data public and retracted its erroneous figure of 520,000 ha.

Notes:

1. "Permerintah Disinyalir Memanipulasi Huas Hutan Yang Terbakar." ["Manipulation of Burnt Area Pointed Out to Government."] Suara Pembaruan, 10 August 1999.
2. Letter from D.M. Nasution, Minister of Forestry and Estate Crops, to the Governor of East Kalimantan, "Re: Clarification of the 1997–98 forest and grasslands fire data," SK 718/Menhutbun-V/1999, July 5, 1999. [Unofficial translation.]
3. Minutes of Meeting with Governor of East Kalimantan, Head of the Provincial Forest Office and German-assisted forest and forest fires project staff, July 8, 1999, Samarinda, East Kalimantan.
4. Statement by Lothar Zimmer, German Federal Ministry for Economic Cooperation and Development, Consultative Group on Indonesia Meeting, Paris, July 28-29, 1999.

IV. COUNTING THE COST: IMPACTS OF THE 1997–98 FIRES

To date there have been three major attempts to value the costs of the 1997-98 fires and haze. One by WWF-Indonesia and the Singapore Economy and Environment Programme for South East Asia (EEPSEA), covering August 1 through October 31, 1997, yielded a figure of nearly $4.5 billion.[59] The Environment Ministry's Environmental Emergency Project (EEP) gave a figure of $2.4 billion.[60] Neither estimate included losses incurred from the 1998 fires.

Building on these earlier studies, a technical assistance team funded by the Asian Development Bank (ADB) and working with the national development planning agency (BAPPENAS) concluded that total losses from the 1997–98 fires and haze were between $8.9 billion and $9.7 billion (based on two sets of assumptions about the value of forest loss), with a mean value of $9.3 billion.[61] These figures are summarized in Table 2; methodological notes are provided in Appendix A.

The WWF-EEPSEA valuation study, which estimated losses of nearly $4.5 billion for 1997 alone, provides some sobering comparisons on the magnitude of losses. It noted that total 1997 damages are:

■ more than the damages assessed for purposes of legal liability in the Exxon Valdez oil spill and the Bhopal (India) chemical release disasters combined;
■ more than the amount needed to provide all of Indonesia's 120 million rural poor with basic sanitation, water, and sewerage services;
■ more than double the total foreign aid to Indonesia annually; and
■ equal to 2.5 percent of Indonesia's gross national product (GNP).

In some cases it is impossible to disaggregate from other factors the losses directly due to the fires and haze. For example, the decline in Indonesia's agricultural production during 1997 and 1998 was caused by drought as well as by fire, while declines in tourism are attributable to the Asian economic crisis and the political unrest in 1998 as well as to the haze.

All three analyses of the costs incurred by the fires and haze were conservative in their assumptions and did not take into account a number of probable but difficult to quantify costs such as long-term health damage, increased risk of cancer, and crop losses due to reduced photosynthesis and pollination. Research by the Malaysian Rubber Board's research institute, for example, indicates that the fires reduced photosynthesis by 10 percent.

Much of the cost of the fire damage probably cannot be estimated. Assigning a dollar value to the destruction of some of the last intact lowland forest in Sumatra, the death of a large percentage of Indonesia's remaining wild orangutans, or the shortened life span of medically vulnerable people made terminally ill by the haze is impossible.

According to one researcher, because the haze of the fires weakened photosynthesis activity, rubber tree growth was affected and the volume of latex produced dropped noticeably.[62] If these less quantifiable costs could be reliably counted, it is likely that they would more than offset distortions caused by the difficulty of disaggregating fire and drought economic losses in the cases of agriculture and tourism. A total economic loss in the range of $8 billion to $10 billion therefore appears to be the most reliable estimate to date, based on current data and methodologies.

Much of the cost of the fire damage probably cannot be estimated. Assigning a dollar value to the destruction of some of the last intact lowland forest in Sumatra, the death of a large percentage of Indonesia's remaining wild orangutans, or the shortened life span of medically vulnerable people made terminally ill by the haze is impossible. But even this conservative and partial assessment of the monetary costs gives policymakers and the people of the region a tangible way of understanding the destruction wrought by the fires and haze.

TABLE 2

The Economic Cost of the 1997–98 Fires and Haze (million U.S. dollars)

SECTOR	Estimated Economic Losses		
	Minimum	Maximum	Mean
AGRICULTURE			
Farm crops	2,431	2,431	2,431
Plantation crops	319	319	319
FORESTRY			
Timber from natural forests (logged and unlogged)	1,461	2,165	1,813
Lost growth in natural forests	256	377	316
Timber from plantations	94	94	94
Nontimber forest products	586	586	586
Flood protection	404	404	404
Erosion and siltation	1,586	1,586	1,586
Carbon sink	1,446	1,446	1,446
HEALTH	145	145	145
TRANSMIGRATION AND BUILDINGS AND PROPERTY	1	1	1
TRANSPORTATION	18	49	33
TOURISM	111	111	111
FIREFIGHTING COSTS	12	11	12
TOTAL	**8,870**	**9,726**	**9,298**

Source: BAPPENAS 1999.

EFFECTS ON FOREST FAUNA AND FLORA

The effects of the 1997–98 fires on the rich forest biodiversity of Kalimantan and Sumatra are largely unknown at this time, for several reasons. First, only a few preliminary field surveys have been carried out in the affected areas. Second, basic understanding about the functioning of rainforest and peat-swamp ecosystems, and about the basic biology of most species in those ecosystems, is extremely limited. Third, it is difficult to disaggregate the impacts of fire from those of drought and land-clearing activities. Finally, the direct effects of fire may be dwarfed in the long term by the indirect impacts of accelerated human occupation and use of formerly forested areas that, having been burned, became more accessible to humans.

What is known about the effects of the most recent fires on wildlife is derived from a few preliminary field assessments and what can be extrapolated from more extensive studies carried out after the 1982–83 Kalimantan fires. These findings are summarized below and are discussed in detail in Appendix B.

Forest vegetation. Where fires are very hot, the soil surface hardens, making it difficult for seeds to sprout and causing nutrient-rich ash to be washed away by the first heavy rain. Damage tends to be greater in logged and otherwise disturbed forests. Peat-swamp forests are particularly vulnerable to fire. Repeated cycles of burning, such as those Indonesia has experienced, can completely transform forest into grassland or scrubland. Apart from the direct effects of the fires, the opening of new areas for cultivation and settlement in previously forested burned areas intensifies degradation of adjacent unburned forest areas by human activities.

Primates. The fires had a particularly severe impact on the orangutan (*Pongo pygmaeus*), particularly in Kalimantan. (*See Map 3.*) Studies after the 1982–83 fires concluded that orangutans were able to alter their diet and recovered fairly well, but the scale of the 1997–98 fires exceeded their ability to adapt. Hundreds of adult orangutans were killed by villagers in Kalimantan as they fled from the drought and fires, and hundreds more orphaned juveniles were captured and sold on the illegal wildlife market. Primatologists believe that the most recent fires mark the beginning of a steep decline in the Borneo orangutan population, which was already dropping due to habitat loss and poaching.

The proboscis monkey (*Nasalis larvatus*), a threatened species found almost exclusively in riverine and coastal habitats, was probably the primate species that lost the greatest percentage of its habitat to the fires. Some other primate species do not appear to be so badly affected in the short term. [63]

Birds. Bird populations were probably seriously affected by the fires. Many birds become disoriented in smoke and fall to the ground, while fruit-eating species such as hornbills lose their source of food. Hornbills disappeared completely from study areas in East Kalimantan after the 1982–83 fires, presumably for this reason. Insect-eating species of birds tend to do well after fires because populations of wood-eating insects increase in response to the enormous supplies of dead wood.

Reptiles and amphibians. With the exception of species that live in relatively deep water, reptiles and amphibians are extremely sensitive to fire and appear to have suffered severe population declines in areas burned during 1997–98.

Insects and invertebrates. Wood-eating species increase after fires. Studies of the 1982–83 Kalimantan fire indicate that butterfly populations also increased and that the number of soil- and litter-dwelling invertebrates recovered within three years. The rich but little-studied invertebrate fauna of the forest canopy, however, is presumably completely destroyed along with the canopy.

Summarizing all the data available by early 1999, the BAPPENAS-ADB study offered the following general conclusions:

■ Drought-assisted fires in 1997–98 caused major biodiversity losses, particularly in Sumatra and Kalimantan, although the vast spread and remoteness of many damaged areas makes it impossible to obtain a precise estimate of losses.

■ The massive areas burned in Kalimantan resulted in large losses to the already shrinking lowland evergreen and semievergreen forests and to swamp and peat forests, with the most serious damage taking place in areas previously subject to logging and fire.

■ Some 17 protected areas were damaged by fire, including the nationally and internationally important Kutai National Park in East Kalimantan, which was severely damaged. Much of the park had previously been logged and burned, making it more susceptible to fire damage in 1998.

■ Deaths of rare and endangered animal species due to the fires was compounded by the hunting of disoriented animals for food and sale and the killing of some that strayed into human settlements.

■ Severe damage was not limited to natural biodiversity but also affected the biodiversity of agricultural ecosystems, including locally evolved cultivars, many of which were probably lost. [64]

EFFECTS ON WATER FLOWS AND WATER QUALITY

As with fauna and flora, the impacts of the 1997–98 fires on water flows and quality have not yet been systematically studied. A good deal is known, however, about the general effects of fire on water flows and quality, and considerable data are available from the 1982–83 fires in East Kalimantan. The main point is that fires greatly increase erosion potential. When a fire episode is followed by a period of heavy rains, the amounts of ash, soil, and vegetative matter carried into water systems increase dramatically.

El Niño-related drought events are sometimes followed by a year of above-average rainfall, called La Niña.[65] This was the case in Indonesia during the rainy seasons of 1983-84 and 1998-99.[66] In 1983-84 the rains resulted in much heavier flooding along East Kalimantan's rivers than would be expected from the amount of rain.[67] The reason was that the fires had impaired the hydrological performance of the forest. In the two previous dry seasons, the fires destroyed ground vegetation and leaf litter (which usually slow surface runoff), hardened the surface of the soil (thus restricting water infiltration), and reduced the capacity of the peat swamps to retain water. Runoff from burned areas carried soil, ash, and woody debris into rivers and lakes, causing heavy sediment loads and biological pollution.[68] (The magnitude of postfire erosion can be staggering. Researchers estimate that in southeastern Australia, a moderate day-long rainfall that occurred a week after a forest fire washed at least 22 metric tons of soil and ash from each hectare of burned-over forest.)[69]

When large sediment loads are washed into streams and rivers, aquatic life is smothered in mud. The sediment brings with it large amounts of nutrients, especially nitrogen and phosphorus, that pollute the water and cause algal blooms. Two years after the 1982–83 fires, nutrient levels in streams draining areas of burned forest in Sabah were twice as high as those of streams draining adjacent unburned forest.[70] Changes in water chemistry after the 1982–83 fires reduced fish populations already affected by a year of below-normal water levels. The incidence of disease in fish rose, and fish catches fell. The endangered Mahakam River dolphin, a marine mammal adapted to living in rivers, suffered an outbreak of disease.[71] The loss of forest cover over streams also exposes the water to direct sunlight, increasing water temperatures to levels that are unfavorable for some fish and aquatic organisms.

EFFECTS ON THE ATMOSPHERE

Most of the gases present in the haze created by the 1997–98 fires play direct or indirect roles in regulating the Earth's atmosphere,[72] and their release made a large contribution to the atmospheric concentration of the greenhouse gases that are generally acknowledged to cause global warming. Several studies have attempted to quantify the volume of biomass that was burned in the 1997 fires in order to estimate accurately the amount of carbon dioxide released from above-ground vegetation.[73] But below-ground burning in peat-swamp forests may in fact be the greater contributor to carbon release because of the special characteristics of peat swamps.

Peat forests cover about 400 million ha of the Earth's surface. About 90 percent of peat swamps lie in temperate and boreal latitudes. The remainder are in the tropics, and Indonesia contains about 60 percent of those tropical peat forests (20 million to 30 million ha). Unlike forests and grasslands, peatlands accumulate carbon over thousands of years and thus represent an important carbon sink. The total global carbon storage capacity of peat is 240 to 480 gigatons (Gt)—about 20 percent of the total global organic carbon store and 200 times more than the amount of carbon released annually from the combustion of fossil fuel and from deforestation.[74]

The total amount of carbon in tropical peat is at least 20 Gt. The ombrogeneous lowland forest peats of Kalimantan (6.8 million ha) and Sumatra (8.3 million ha) store the highest amounts of carbon per hectare in the world—10 times that of tropical forest biomass.[75] The costs of carbon storage via afforestation are estimated to be in the range of $3 to $4 per ton of carbon. Assuming that the average depth of Indonesian peat is 5 meters (m), the carbon store could be as high as 2,500 tons of carbon per hectare. The value of these peatlands as a carbon sink—in the framework of the Clean Development Mechanism (CDM) being developed under the Kyoto Protocol to the Framework Convention on Climate Change, for example—is obviously very high, ranging from $3,600 per ha to as much as $18,000 per ha for deep peats.[76]

Burning peat releases vast amounts of carbon that has been stored for thousands of years. Smoke from peat fires contains high levels of sulfur oxides. The BAPPENAS-ADB study estimated that the total amount of

carbon released into the atmosphere during the 1997-98 fires was 206.6 million tons, over 75 percent of which was derived from the combustion of peat. Unlike the case of above-ground vegetation, carbon emissions due to peat fires are not offset over time by vegetation regrowth. In mid-1999 the World Bank claimed that

"Indonesia's fires in 1997 were estimated to have contributed about 30 percent of all man-made carbon emissions globally—more than the entire emissions from man-made sources from North America."[77]

EFFECTS ON HUMAN HEALTH

Anyone unfortunate enough to have suffered in the choking gloom that enveloped the region intuitively understands that the smog was bad for health. A resident of Sumatra's Jambi province told a reporter, "This morning, like most mornings, I woke with a headache. In my stomach I feel very strange, and my eyes, they sting. Jambi cannot handle these things. This has gone on too long. We have not seen the sun for more than a month. We are suffocating."[78] Assessing the health impacts in a systematic and quantitative way, however, presents serious methodological problems.

The severity of effects on human health from a "haze episode" such as that caused by the 1997–98 fires depends on the level of the constituent pollutants and the length of exposure of the population to them. (*See Box 8.*) It is not possible at this writing to give a comprehensive and reliable account of how many people were exposed to what pollutants, at what level, and for how long during the 1997–98 fires. Without those data, it is not possible to accurately assess the magnitude of health impacts. Nevertheless, the data that do exist provide a sobering snapshot of the likely health consequences.

Indonesia. Unfortunately, the spottiest and least reliable data come from Indonesia, where the haze was heaviest and the most people were affected. Air quality monitoring in Indonesia is carried out by three government agencies, independent of each other.[79] The 10 Indonesian government stations in Sumatra and Kalimantan that monitor air quality—like those throughout the country—only sample for total suspended particles

(TSP), which range up to 35 or 50 micrometers (μm) in diameter and thus cannot be used directly to assess respiratory concentration of particles.[80] Heil (1998) reports that a field check of Indonesian sampling methods revealed various methodological flaws that significantly bias TSP sampling results downward. She concludes, however, that the data still "reflect approximately the range of concentrations that occurred during the 1997 Indonesian haze."

It is possible to estimate levels of PM_{10} (particles with a diameter of less than 10 μm) in Indonesia from general data on TSP levels, based on sampling done at locations in neighboring countries where both types of data were collected during the fires. Samples taken in Malaysia indicate TSP/PM_{10} ratios of 70 to over 80 percent. When particles are transported long distances, the heavier ones tend to fall out of the mix, elevating PM_{10} levels. Still, Ferrari (1997) concludes that if the Indonesian TSP samples were taken at locations "30–50 km from the source of the fires, then it may be appropriate to assume that TSP values would translate to PM_{10} levels of at least 50 percent."

---

8 HUMAN HEALTH EFFECTS OF SMOKE FROM FOREST FIRES

Emissions from forest fires are a complex mixture of solid, liquid, and gaseous compounds, and their composition varies depending on the chemical composition of the burning biomass and the conditions and efficiency of combustion. Forest fires produce gaseous compounds, including carbon monoxide, sulfur dioxide, methane, nitrogen oxides, and various organic compounds. From a human health perspective, the most important component of smoke from forest fires consists of suspended particles (a combination of solids and liquids), mainly composed of organic and elemental carbon.[1]

Of particular concern are particles with a diameter of less than 10 micrometers (μm), termed PM_{10}. When people inhale particulate matter, particles are retained in various parts of the respiratory system according to their size. Particles over 10 μm in diameter come to rest in the nose, throat, and larynx and remain there only for several hours. Particles below 10 μm in diameter are able to advance into

the thoracic region (chest), generally come to rest in the trachea-bronchial area, and are removed over several hours to a day. Finer particles, below 6 μm, penetrate into the air cells and passages of the lungs (alveolae), and their elimination takes from days to years.[2] Particles below 2.5 μm ($PM_{2.5}$) have the most serious and long-lasting effects because they can most easily reach the lower regions of the lungs.[3] Total particulate matter (TPM) emitted from forest fires with flaming combustion contains 80 to 95 percent fine particles ($PM_{2.5}$); that from smouldering combustion contains from 90 to nearly 100 percent.[4]

The health effects of breathing particulate matter depend not only on the size of the particles but also on the nature of the toxic compounds adsorbed (gathered in a condensed layer) on their surface. In the case of smoke from forest fires, a class of more than 100 compounds called polycyclic aromatic hydrocarbons (PAHs), many of which are known to be carcinogenic, are of particular concern.

PAHs are formed when combustion is incomplete because of an insufficient supply of oxygen. Burning of wood and charcoal yields a higher level of PAHs than does combustion of gas, petroleum, or coal. PAHs tend to adsorb on particles of 10 μm or less, and thus they penetrate deeply into the lungs along with the particles.[5]

Elevated levels of PM_{10} particles in the ambient air, especially when a significant proportion of $PM_{2.5}$ particles is present, are associated with an increase in acute health hazards ranging from "acute respiratory symptoms and illness including bronchitis, asthma, pneumonia and upper respiratory infection, impaired lung function, hospitalization for respiratory and cardiac disease to increases in mortality. The organic constituents have been shown to induce some inflammations and suppress the defense capability toward bacteria"[6]

The United States has set standards of 150 micrograms (μg) per m^3 for PM_{10} and 65 $μg/m^3$ $PM_{2.5}$. Concentrations above these levels are considered unhealthful or,

at higher levels, hazardous. At particular risk are children, the elderly, and those with preexisting conditions such as asthma and heart disease.[7] Given the well-known carcinogenic properties of PAHs and the potential long-term effects on lung and heart function of extended exposure to them, it can be assumed that long-term negative health effects are a probable outcome of elevated levels of exposure to these pollutants. Malaysia and Singapore use the same standard for PM_{10} as does the United States (150 $μg/m^3$). Indonesia only has a standard for total particulate matter (TPM): 260 $μg/m^3$ in a 24-hour period.

Notes:

1. Heil, 1998.
2. Ibid.
3. USEPA, 1998.
4. USDA, 1997.
5. Heil, 1998.
6. Heil, 1998: 4.
7. USEPA, 1998.

Using that ratio—PM$_{10}$ levels at 50 percent of TSP—the figures recorded for Indonesia in September-November 1997 are extremely high. In late September, TSP levels of 4,000 µg/m³ and above were reported for Jambi and Central Kalimantan provinces, and another peak of around 3,500 µg/m³ was reached in mid-October. Levels in other provinces were not as high, but they nevertheless rose at least an order of magnitude above Indonesia's maximum 24-hour TSP standard of 260 µg/m³. [81]

Several efforts to directly monitor PM$_{10}$ levels in Indonesia were carried out during the fires:

■ A short-term study conducted by BAPEDAL in Jambi province during October 3–5, 1997, reported PM$_{10}$ levels of more than 1,000 µg/m³—nearly seven times the maximum level considered safe by the United States. [82]

■ The U.S. Environmental Protection Agency monitored particulate matter (PM) at two haze-affected sites in Sumatra between November 4 and 8, 1997. Levels of both PM$_{10}$ and PM$_{2.5}$ exceeded U.S. national ambient air quality standards (NAAQS) by large margins on all five days, with values on several days reaching hazardous levels. Most of the particles were PM$_{2.5}$, and "the specific composition of the particles and presence of particular PAH compounds are characteristic of wood smoke." [83]

These sketchy data probably underestimate pollution levels experienced in some provinces because monitoring stations are located in provincial capitals, some of which are not representative of the province's experience with the haze. [84]

Data on the health impacts of the haze in Indonesia are sparse and are based only on 1997 observations. A report from West Kalimantan's provincial health office noted significant increases in respiratory diseases in the capital city of Pontianak. [85] Heil (1998: 13) observes that "health statistics registered a considerable increase of upper respiratory infection, asthma, bronchitis and pneumonia, as well as eye and skin irritation. Besides the physical effects, depression and anxiety syndromes occurred more frequently. . . .The persistence of exceedingly high particle levels leads to an overload of deposited particles within the respiratory system, which is most likely to induce chronic, long-term respiratory diseases."

The Environment Ministry's fire and haze study estimated, on the basis of data from eight affected provinces, that about 12.3 million people were exposed to the haze during September-November 1997. Of these, 527 died, nearly 16,000 were hospitalized, and more than 36,000 received some form of outpatient treatment. (*See Table 3.*)

While data from hospital admissions and outpatient visits to doctors indicate significant health effects from the haze, these figures probably underreport the actual impacts by an order of magnitude, for three reasons. First, unlike the situation in Malaysia and Singapore, "almost no information was given to the public concerning the level of air pollution and the ensuing health effects." [86] Second, because of the difficulties of access, cost, and the "chronic inadequacy of services," [87] Indonesians living in rural areas are unlikely to visit a hospital or health clinic unless they are suffering acute symptoms of illness. Third, Indonesians, particularly those living in rural areas of islands such as Kalimantan and Sumatra, depend on traditional healers and herbal medicines for a great deal of their health care. [88]

TABLE 3

Health Effects from Fire-Related Haze Exposure in Eight Indonesian Provinces, September–November 1997

HEALTH EFFECTS	NUMBER OF CASES
Death	527
Asthma	298,125
Bronchitis	58,095
Acute respiratory infection (ARI)	1,446,120
Daily activity constraint (number of days)	4,758,600
Increase in outpatient treatments	36,462
Increase in hospitalizations	15,822
Lost work days	2,446,352

Note: The provinces studied were Riau, West Sumatra, Jambi, South Sumatra, West Kalimantan, Central Kalimantan, South Kalimantan, and East Kalimantan.

Source: State Ministry for Environment and UNDP, 1998.

Faced with chronic (rather than acute) symptoms, such as respiratory illness or skin and eye irritation, many rural Indonesians simply will not visit a doctor. That leaves only the inferences that can be drawn from the fact that as many as 21 million people spent months breathing undetermined but extremely high levels of pollutants known to cause both acute and long-term health effects. (The figure 21 million is the number of people living in 1995 in the six provinces most affected by haze from the 1997 fires—South Sumatra, Riau, Jambi, Central Kalimantan, South Kalimantan, and West Kalimantan.)

> *As many as 21 million people spent months breathing undetermined but extremely high levels of pollutants known to cause both acute and long-term health effects.*

Not all of the health effects of the haze were caused by breathing polluted air; for example 26 people were killed on a river in South Kalimantan when a water taxi collided with a tugboat towing a coal barge in visibility reduced to less than 5 m by the haze.[89] In North Sumatra the haze was suspected to have been at least partly responsible for the deaths of 234 people in the September 1997 crash of an Indonesian Airbus near the city of Medan. Haze from the fires was reported to be thick at the time in the area, and the Medan airport closed shortly after the crash because of poor visibility. Aviation experts noted that thick smoke could also have caused the crash by creating turbulence or cutting oxygen flow to the engines.[90]

Malaysia, Singapore, and Thailand. Haze from the 1997 fires raised air pollution in Malaysia to alarming levels. In late September, the API reached values of over 800 in Kuching, Sarawak, and 300 in Kuala Lumpur. API values of 101 to 200 are categorized as "unhealthy," values of 201 to 300 as "very unhealthy," and values of 301 to 500 as "hazardous." Values of 501 or more are considered to pose risks of "significant harm." These high values were caused by elevated levels of suspended particles in the air.[91] Normal nonhaze PM$_{10}$ levels are about 50 $\mu g/m^3$ in Kuching and 65 $\mu g/m^3$ in Kuala Lumpur. At the height of the haze these levels went as high as 931 $\mu g/m^3$ in Kuching and 421 $\mu g/m^3$ in Kuala Lumpur.[92]

In Sarawak, as the smog reduced visibility to a meter or less, a state of emergency was declared on September 19, closing factories and schools. The press reported that at least 5,000 people had sought medical help for haze-related complaints; one report noted that 3,000 people had sought treatment on a single day. At one point the government began to make contingency plans for evacuating all 2 million residents of the state.[93]

A World Health Organization consultant, assessing the health impacts of the haze, noted that the limited availability of baseline morbidity and mortality information in Malaysia makes quantitative assessment difficult. However, the Ministry of Health monitored the incidence of acute respiratory infections, asthma, and conjunctivitis at three hospitals in Peninsular Malaysia during August and September, and similar data were collected in Sarawak during September. Comparison of these data with same-day API data showed a clear relationship between the incidence of illness and concentrations of inhalable particulates (PM$_{10}$) at hospitals in the Kuala Lumpur area and for the state of Sarawak.[94] In early October, Malaysia's deputy health minister told members of Parliament that there had been a 65-percent increase in asthma cases recently and that conjunctivitis cases had risen by 61 percent among adults and by 44 percent among children.[95]

Singapore, at the southern tip of the Malay Peninsula, also experienced high levels of haze-related pollution, although the situation was not as extreme as in Sarawak and in much of Indonesia. Monthly PM$_{10}$ values, which usually fluctuate between 30 and 50 $\mu g/m^3$, increased to between 60 and 110 $\mu g/m^3$ during September-October 1997. The incidence of medical complaints related to the haze rose about 30 percent during this period, but there was no significant increase in hospital admissions or mortality, indicating that the short-term health effects in Singapore were relatively mild. Of concern for the longer term was the finding that 94 percent of the particles in the haze in Singapore were below 2.5 μm in diameter—that is, they were the fine particles that settle deepest in the lungs and take the longest time to eliminate.[96]

One measure of the severity of the haze was its effects on southern Thailand, which lies some 1,200 km from the nearest major clusters of the 1997 fires in Kalimantan and some 800 km from the major fire areas in Sumatra. Although PM$_{10}$ levels only reached an average high of 69 $\mu g/m^3$ during September, analysis of health data for the period "showed elevated and widespread short-term respiratory and cardiovascular health effects in the same period. Approximately 20,000 visits and 1,000 admitted cases are estimated as a lower end of excess health effects from the 1997 haze in southern Thailand. The number may increase or double if visits or admissions in private health facilities are included."[97]

V. DIRECT MEASURES TO COUNTER FUTURE FIRE OUTBREAKS: RECOMMENDATIONS

As stated at the outset, the most effective policy reform measures that Indonesia can take to prevent another fire catastrophe during the next El Niño-induced drought are essentially congruent with the broader forest policy reform agenda discussed in Part II of this report. This does not mean that incremental and technical steps to deal with the proximate causes of forest and brush fires in Indonesia should not be taken; such steps are urgently needed, and must be integrated into the larger forest reform agenda. As Vayda (1998) has noted:

There is a tendency . . . certainly present among academics, policy analysts, and environmentalists, to highlight allegedly deeper or underlying causes and to be dismissive towards proximate and, especially, so-called accidental causes. . . . While it may be fairly argued that we need a 30-year perspective to understand why, under the Suharto regime, vast areas of Indonesia's primary rainforests were degraded and thus became more fire-susceptible, future fire prevention will of course need to take place not in the rainforests as they were prior to three decades of predatory resource exploitation but rather in a world of degraded and more fire-susceptible forests [which argues for] serious attention to the proximate causes. . . .These two arguments are not mutually exclusive, and policymakers, environmentalists, foresters, and other concerned people would do well to heed both.

Following this advice, this chapter recommends steps directly related to fire prediction and suppression that the government and its supporting donor agencies can and should carry out immediately. [98] Part II then moves on to the broader forest policy reform agenda that is the context for effective fire control.

■ Study and learn from the 1997–98 fires.

Although a commendable effort was made to study the impacts of the 1982–83 fires, many of the resulting policy and procedural recommendations were not adopted by the government, and those that were adopted were generally not adequately funded or implemented. More important, causal factors were not thoroughly analyzed, especially the attitudes and motivations of the major groups thought to be responsible for the fires. Some management needs and fire causes are already documented well enough to guide initial government actions. It is critical, however, that data analysis and field investigations be continued and that the government incorporate the findings into policy. The following questions have been partially answered in this and other reports but merit further investigation:

■ How much of each vegetation type burned?
■ What are the relationships among fire occurrence, vegetation type, land use, and land ownership? Specifically, what constellation of economic incentives, property rights, land and forest exploitation practices, and human settlement patterns was most conducive to the ignition and spread of the fires?
■ What human actions directly caused or exacerbated the fires; what categories of people or firms were linked to each type of action; and what were their motivations?
■ What were the most serious effects on human health, forest ecosystems, and wildlife, and what are the likely long-term impacts?
■ What specific institutional and political factors led to the ineffective government effort to prevent and control the fires?
■ What were the roles of drought and other natural factors in raising fire hazard, and to what extent were these factors predicted and monitored? Would better forecasting have mattered?

■ Establish coordinated and flexible institutional mechanisms for fire prevention and suppression.

Government responses to the 1997–98 fires were poorly coordinated both vertically and horizontally, leading to serious ineffectiveness. At the central level, the government needs to appoint one agency as "fire czar," with the authority to coordinate and compel action from other agencies, particularly when a fire emergency has been declared. A similar structure needs to be created at the provincial and district levels, especially in areas identified as having high levels of fire hazard and risk. Communications between levels of government, in technical as well as organizational terms, need to be modernized and streamlined. [99]

NGOs played an important role in both monitoring fire impacts and distributing assistance to fire and haze victims during 1997–98, and at least some government agencies accepted NGOs as full partners in dealing with the disaster. With the fire emergency ended for the time being, government and donor efforts have gone back to "business as usual," with NGOs left out of decisionmaking processes, except for token "consultations" from time to time. The valuable role of NGOs should be more formally recognized by the government and specifically funded by donor agencies.

■ Identify high fire hazard areas.

Fire hazard is a measure of the amount, type, and dryness of potential fuel in an area. There are two steps in identifying high fire hazard areas: assessment of the amount and type of potential fuel, and assessment (and prediction) of drought. A fire danger rating system for monitoring fire hazard has been developed by the German-funded Integrated Forest Fire Management Project and is already in place in East Kalimantan. [100] The system proved its effectiveness in 1997, and other provinces prone to wildfires should set up a similar system as soon as possible. [101]

Areas where there is likely to be a high level of potential fuel include logged-over concession areas, newly cleared timber and oil palm plantations, and transmigration sites. As these are also areas of high fire risk, they must receive special attention in monitoring hazard levels.

The science for predicting major El Niño-related droughts has developed to the point where these events can be foreseen months in advance,[102] as was true for Indonesia in early 1997. Local drought conditions can vary considerably however and need to be monitored in tandem with potential fuel loads.

■ Reduce fire hazard by reducing logging and land-clearing waste.

Fire hazard must be not only monitored but also reduced. Drought cannot be mitigated, but quantities of potential fuel can certainly be decreased. On logging concessions, the government needs to demand and enforce better logging practices and to change the current system of taxing log production to one that will provide incentives to minimize the amount of waste wood left behind by logging operations. A great deal of illegal logging occurs on logging concessions. Making concession holders legally and financially accountable for suppressing illegal logging on their concessions would help minimize fuel loads.

On oil palm, timber, and pulp plantations, the government should phase in a requirement that plantations be established (and rotated) using the "zero-burning" techniques long utilized in Malaysia and accepted in principle by the Indonesian government.[103] Indonesia is already developing a system to certify logging operations, as discussed in Part II. Certification of responsibly produced palm oil and paper products would provide an additional incentive for firms to find alternatives to burning. Plantation firms must also be made legally and financially responsible for the actions of contractors hired to clear their lands. And a strong policy commitment to siting plantations only on already degraded land would help as well.

Regulatory approaches can only go so far, however. Plantation firms need concrete economic incentives to reduce potential fuel loads, which will happen only when there are uses for vegetative waste more financially attractive than the savings derived by burning them. A 1998 study by the World Bank's Economic Development Institute predicts that a wood-chip market "will soon emerge in both Sumatra and Borneo due to new and expanded pulpmill operations, which require wood supply in excess of existing plantation establishment and growth rates." The study notes that two large pulp and paper mills in Sumatra are already utilizing "virtually all of the wood from land clearing."[104] Increased incentives for this kind of waste wood utilization, combined with a strong "zero-burning" policy and the phaseout of current policies allowing pulp mills to cut surrounding forest for feedstock until their plantations begin to produce, could change firms' economic calculus and significantly reduce fuel loads.

In addition, technologies exist in the United States and other industrial countries for converting wood residues from land clearing into mulch used as fertilizer, nursery potting soil, and a soil enhancement medium for land reclamation.[105] These technologies have not been tested in Indonesia and would probably require initial subsidies to get established. This would be a good area to integrate into one or more of the many donor agencies' initiatives on fire control.

■ Reduce fire risk.

Fire risk is the measure of the probability that fuel will ignite. The level of risk is usually related to negligent or deliberate human action, and reducing fire risk is therefore a matter of managing people and institutions, not managing fire. The total exclusion of fire from natural vegetation and agricultural areas would impose economic hardship, could disrupt natural ecological processes, and is impossible to achieve, in any case. But currently, the use of fire to clear vegetation and dispose of agricultural wastes is virtually unregulated in Indonesia.

The circumstances under which fire may be legitimately used for these purposes need to be clearly defined in law and policy. Within that framework, Indonesia needs to introduce and strictly enforce a burning permit system. Permits should specify the area to be burned, the term for which the permit is valid, requirements for firebreaks and other control measures, and the identity of the responsible owner or manager of the land. During periods of high fire hazard, permits should not be issued, and any burning during that period should be presumed to be in violation of the law.[106]

Fire education and awareness campaigns are greatly needed in Indonesia. Both smallholders and commercial operators need to be taught proper methods for conducting controlled burns. (Traditional shifting cultivators are already skilled in this regard, and their techniques might be transferable to the majority of Indonesia's shifting cultivators, who are not from such traditional cultures.) Indonesia, like other countries, requires people who wish to acquire drivers' licenses to know how to drive safely. At least the same level of care should be exercised in granting people permission to set fires.

Children need to be targeted as well, through the school system. The careless use of fire is a learned behavior, transmitted from generation to generation. Fire prevention awareness campaigns in the schools are an important step in ensuring that the next generation has a greater appreciation for the damage that fire can do and a greater knowledge of how to use it safely.[107]

Although increasing the certainty and severity of criminal punishment for unlawful use of fire is not a total solution to reducing fire risk, it should be part of the government's approach. The laws are probably strict enough; the problems lie in detecting violations and obtaining convictions in court. Detection of violations can be improved through strengthened government fire-monitoring programs, both on the ground and via aerial and remote-sensing methods, and through independent citizen forest-monitoring efforts (discussed in Part II).

The problems in proving liability that prevent cases from being successfully tried in court point to a need for legal and procedural changes. The Regional Environmental Impact Assessment Agency (BAPEDALDA) in East Kalimantan, for example, obtained strong evidence of unlawful burning against three oil palm plantation firms in early 1998, including extensive eyewitness reports and photographic evidence. But, in May 1998, the provincial police made it clear that they would drop their investigation of the firms, ostensibly because there was not enough proof to take them to court.[108] Whether this was an evidentiary problem or a simple case of collusion between the police and the firms is unclear.[109]

■ **Monitor and mitigate health impacts.**

As noted, Indonesian capacity to monitor fire-induced air pollution levels proved to be extremely weak in 1997–98. It needs to be substantially upgraded over the next few years. Relatively simple and inexpensive technologies exist for monitoring the levels and composition of pollution from fires. Monitoring capacity should be developed particularly in those areas that were shown to be at greatest risk of high pollution levels in 1997–98. Monitoring, however, requires much more than wider dissemination and use of monitoring devices. Monitoring stations need to be staffed with well-trained technicians, and systems need to be put in place to ensure that data can be rapidly collected, analyzed, and made available to relevant agencies, the press, and the public. Hospitals and health clinics in high-risk areas should ensure that their staff keep accurate records of haze-related complaints and admissions.

If this information is made available during periods of fire-induced pollution, people will be able to take action to protect themselves. They need to be informed, however, on what actions to take. In 1997–98 a great deal of attention was focused on the distribution of ordinary surgical masks, but these do not block inhalation of the fine particles that are of most concern, and the kind of mask that does block such particles is probably too expensive for mass public use in Indonesia. Simple measures such as remaining indoors and avoiding strenuous exertion are more realistic self-help options.[110]

Since haze greatly reduces visibility, risks of transportation accidents rise dramatically, as was the case in 1997–98. During such times, the public—and public transportation operators—should be encouraged to avoid all unnecessary travel and to take extra precautions when driving or when piloting river and seagoing vessels.

■ **Freeze issuance of "salvage felling" permits for burned-over forest areas until the government can effectively monitor and enforce implementation of permit provisions and restrictions.**

Indonesian regulations allow logging firms to conduct limited "salvage felling" in burned-over forest areas. Following on the 1998 fires in East Kalimantan, however, logging companies have blatantly abused these regulations, using the "salvage felling" exception to cut large areas of undamaged forest rather than to conduct a final cut of truly damaged areas. Despite policy statements to the contrary, the government has done nothing to enforce these salvage regulations. At the same time, there are in fact more than 1 million ha of severely damaged forests in East Kalimantan where no action at all has been taken, whether salvage felling or rehabilitation efforts.[111] Until effective monitoring and enforcement can be carried out, issuance of "salvage felling" permits should be stopped, and logging firms should be prohibited from conducting such operations.

Those who do not learn from history, it is often said, are condemned to repeat it. Twenty El Niño episodes have occurred since 1877. There is currently much debate about whether the frequency and intensity of these episodes are increasing. A comprehensive statistical analysis conducted in 1997[112] concluded that the tendency toward more El Niño events—and related droughts in Indonesia—since the late 1970s is highly unusual and is unlikely to be accounted for solely by natural variability. Prudence dictates that policymakers should assume that another severe El Niño-triggered drought will occur in Indonesia within the next several years and should act accordingly.

> *Despite Indonesian decrees against the use of fire to clear land and international principles against causing environmental harm to neighboring countries, little has changed . . . The fires and haze have their roots in cronyism and nepotism amongst corporate citizens—the same problems that continue to beset relations between government and business in Indonesia and which contributed to Mr. Suharto's downfall. Laws and decrees against the use of fires to clear land are on the books, but remain unenforced.*
>
> Simon S.C. Tay,
> Chairman
> Singapore Institute of International Affairs
> *International Herald Tribune*, August 31, 1999

Each El Niño-related fire episode from 1982–83 to the present has triggered calls for government action. In recent years, foreign governments and multilateral donor agencies have provided substantial financial and technical assistance for fire monitoring, prevention, and suppression.[113] This assistance had little or no effect on the severity of the 1997–98 fires because the Indonesian government failed to heed advice or take measures to reduce fire risk and hazard through improved land management and agricultural practices. The renewed outbreak of fires in mid-1999—and the government's failure once again to respond to them effectively—shows that little has changed in this regard.

Most Indonesians, governments and residents of neighboring countries, and the global community now believe that rapid and firm action must be taken to control the indiscriminate and careless use of fire in Indonesia, reduce anthropogenic fire hazard and risk factors, and improve the nation's ability to manage fire by developing appropriate policies and laws for prevention, preparedness, and suppression.

Given the dramatic negative impacts of the 1997–98 fires on neighboring countries, these matters can no longer be considered purely domestic policy issues. Even ASEAN—legendarily cautious about "interference" in member-states' "internal affairs"—has systematically taken up the subject of Indonesia's forest fires.[114] When the deliberate actions, or negligent inaction, of one government poses a serious threat to the health and welfare of citizens in neighboring countries, both the governments of those countries and the international community at large have the right and duty to demand action.

If the government of Indonesia takes action forthrightly, it can and should expect the continued financial and political support of the international community. If the government does not do so, and the forests of Indonesia repeatedly burst into flames, polluting the air and blotting out the sun across Southeast Asia in coming years, Indonesia must expect anger greater than that occasioned by the fires of 1997–98 and the possibility of substantial international political and economic sanctions.[115]

NOTES FOR PART I

1 "Indonesian haze hits hazardous levels." AFP, August 4, 1999.

2 "Haze from Indonesian forest and ground fires creeps over south Borneo," AFP, September 20, 1999.

3 Forest Fire Prevention and Control Project, "Current Sumatra Fire Situation," updated August 31, 1999. Available online at: http://www.mdp.co.id/ffpcp/overvw2.htm; "Tahun 2000 hutan Sumatera akan kembali terbakar." ["Sumatra's forests will burn again in 2000."] Republika, July 28, 1999.

4 In Transparency International's 1998 Corruption Perception Index (CPI), a "poll of polls" on perceptions of the level of corruption in 85 countries, Indonesia ranked 80th, nearly as low as Nigeria and Tanzania, and only marginally ahead of the three lowest-ranking countries, Honduras, Paraguay, and Cameroon. The CPI for Indonesia was based on a composite of 10 separate polls of businessmen concerning their perceptions of corruption. For further information, see http://www.transparency.de/documents/cpi/index.html.

5 In the early 1990s, for example, the World Bank was pressing for a number of timber concession management reforms, including performance bonds, auctioning of concessions, changes in the ways stumpage fees are paid, and other measures. At the time, the Forestry Ministry, flush with cash and investments, resisted these reforms and essentially ended the World Bank's forest sector work in the midst of the preparation of a large forest sector loan, the third in a series. In negotiating the International Monetary Fund–World Bank–Asian Development Bank $43 billion economic bailout package during the first part of 1998, however, the government essentially committed to implementing much of the World Bank's long-standing reform agenda as a condition of the bailout package.

6 The Indonesian government has adopted the term "forest and land fires" for the type of fires that ravaged the country in 1997–98 to emphasize the fact that many of the fires actually occur in areas that have been cleared of forest vegetation. Indeed, it is this process of agricultural clearing that is a major cause of the fires. For purposes of this report, "forest fires" should be taken to mean fires occurring either in forests or in areas recently cleared of forest vegetation for agricultural purposes.

7 Goldammer and Siebert, 1990.

8 Verstappen, 1980.

9 Kershaw, 1994.

10 Michielsen, 1882.

11 Whitmore, 1990.

12 Giesen, 1996.

13 Bruenig, 1996.

14 Johns, 1989.

15 Schindler, Thoma, and Panzer, 1989: 68–70.

16 Ibid.: 70.

17 Ibid.

18 Ibid.: 75.

19 Mackie, 1984.

20 Vayda, Colfer, and Brotokusomo, 1980.

21 Pangestu, 1989: 155.

22 Schindler, Thoma, and Panzer, 1989.

23 Ibid.: 113.

24 Ibid.: 113.

25 BAPPENAS, 1999.

26 State Ministry for Environment and UNDP, 1998.

27 "An Act of God," The Economist, July 19, 1997: 77–79.

28 State Ministry for Environment and UNDP, 1998.

29 "Pemda Riau Membiarkan Kasus Pembakaran Hutan" [Riau provincial government ignoring forest arsonist cases], Republika, July 29, 1999.

30 BAPPENAS, 1999.

31 Ibid.: 91.

32 David Wall, personal communication, 1998.

33 ASEAN, 1997.

34 The rural impacts of the 1997 currency devaluation varied a great deal between areas. In regions where export commodities constitute a significant proportion of the local economy, farmers received a windfall from the devaluation. In areas where this was not the case, however, the rising prices had severe negative economic impacts. See Poppele, Sumarto, and Pritchett, 1999.

35 "Borneo Ablaze," BBC, February 20, 1998.

36 "Renewed Indonesia Fires Worry Southeast Asia," International Herald Tribune, February 13, 1998.

37 "Smog decends as fires rage again," The Age, February 22, 1998.

38 International Herald Tribune, April 22, 1998.

39 Kompas Online, April 18, 1998.

40 "Minister: Fires Out of Control in Indonesia," Reuters, April 14, 1998.

41 "Forest Fires Low Priority Says Minister," South China Morning Post, April 21, 1998.

42 "Forests Die as Borneo Prays for Rain: Drought Has Turned Forest into Tinder," International Herald Tribune, April 20, 1998.

43 "Dry season boosts smog fear in S.E. Asia," Reuters, May 22, 1998.

44 The European Union-funded Forest Fire Prevention and Control Project picked up 294 hot spots in the provinces of North Sumatra, Jambi, Riau, and South Sumatra on November 24, 1998. The German-funded Integrated Forest Fire Management Project detected 41 hot spots in East Kalimantan on October 23, 1998.

45 "KC Hopes to Identify Cause of Sudden Haze," Straits Times, December 1, 1998.

46 "Sultan's Brunei Palace Invisible in Haze," Associated Press, April 11, 1998.

47 "Malaysian TV News Cautioned to Stay Quiet on Haze," Associated Press, April 21, 1998.

48 "Haze: Air Quality an 'Official Secret.'" South China Morning Post, August 7, 1999.

49 Ferrari, 1997.

50 "Wound in the World," Asiaweek, July 13, 1984, p. 43.

51 "Bob Hasan: 'Bodoh, Pengusaha Hutan Membakar Hutan'" [Bob Hasan: 'It would be stupid for loggers to burn the forest'], Kompas, October 10, 1997.

52 Vidal, 1997.

53 "Perusahaan Pembakar Hutan Jadi 176," ["176 Firms Have Set Forest Fires"] Republika, September 18, 1997.

54 "Dephut suda Cabut 166 IPK," ["Forestry Ministry Cancels 166 Timber Cutting Licenses"] Suara Pembaruan, October 8, 1997.

55 "Kebakaran Makin Hebat Penanganan Tetap Sama," ["The Fires Increase but Response Remains the Same"] Kompas, October 15, 1997.

56 The meeting was attended by the ministers of information, forestry, agriculture, transmigration, public works, and transportation.

57 "Environment Agency Denies El Niño Responsible for Fires," Indonesian Observer, November 13, 1997.

58 Dennis, 1998.

59 WWF Indonesia Programme and EEPSEA, 1998.

60 State Ministry for Environment and UNDP, 1998.

61 BAPPENAS, 1999.

62 "Smog Cut Sun to Malaysia Rubber," Reuters, September 25, 1998.

63 Yeager and Fredriksson, 1999.

64 BAPPENAS, 1999.

65 Nicholls, 1993.

66 "Rains Bring Hope, Concern to Indonesian Commodities," Reuters, October 1, 1998.

67 Wirawan, 1993.

68 Boer, 1989.

69 Leitch, Flinn, and van de Graaff, 1983.

70 Grip, 1986.

71 Wirawan, 1993.

72 Wirawan, 1993.

73 Liew and others, 1998; Ramon and Wall, 1998.

74 Rieley and Page, 1997.

75 Ombrogeneous peats are "true" peat swamps with an organic accumulation greater than 50 cm and receiving their water and nutrient supply from aerial deposition only. Almost all of the Indonesian peats are ombrogeneous; "topogenous [<50 cm and receiving nutrients from river flow] peat occurs only in a few isolated locations and is comparatively insignificant" (Radjagukguk, 1997).

76 Diemont and others, 1997.

77 World Bank, 1999a.

78 " 'We Are Suffocating': No Escape for the People of Jambi," *Asiaweek*, October 10, 1997, p. 38.

79 The Meteorological and Geophysical Agency (BMG) monitors total particulate matter (TPM) every sixth day; the provincial departments of the Ministry of Health monitor TPM, sulfur dioxide, nitrogen oxides, ozone, and carbon dioxide at weekly intervals; and the Ministry of Social Welfare measures TPM, sulfur dioxide, nitrogen oxides, ozone, and carbon monoxide irregularly (Heil, 1998).

80 Ferrari, 1997.

81 Heil, 1998.

82 Ferrari, 1997.

83 Ferrari, 1997.

84 For example, South Kalimantan's capital, Banjarmasin, sits on the south coast, where prevailing winds in September–November 1997 blow from the sea. In October the city was relatively clear during a visit by one of the authors, while only 40 km away in the interior the haze was so thick that the sun was completely blotted out at midday and visibility was reduced to 30 m or less.

85 Cited in National Institute of Health Research and Development, 1998.

86 Ibid.: 13.

87 Iskandar, 1997: 205–31, 221.

88 Zuhud and Haryanto, 1994; De Beer and McDermott, 1996: 59–62.

89 "26 Found Dead in S. Kalimantan Boat Crash," *Jakarta Post*, October 22, 1997.

90 "Airbus Crash on Sumatra Kills 234," *International Herald Tribune*, September 27–28, 1997.

91 USEPA, 1998.

92 Brauer, 1997.

93 "Heavy Smog in Malaysia Puts Outside Off-Limits," *International Herald Tribune*, September 20–21, 1997; "On Borneo, Anger at Lack of Haze Help," *International Herald Tribune*, September 27–28, 1997; "Malaysia's Kuching Blinded as Smog Rises Sharply," *Reuters*, September 23, 1997.

94 Ibid.

95 "Under the Haze, Illness Rises," *International Herald Tribune*, October 7, 1997.

96 Singapore, 1998.

97 Phonboon and others, 1998: 35.

98 For more detailed technical recommendations on preventing and suppressing forest and land fires, see State Ministry for Environment, 1998; BAPPENAS, 1999.

99 For details on how communications systems for fire prevention and suppression can be improved, see Hansen, 1997.

100 Pickford, 1995; Ridder, 1995.

101 Bird, 1997.

102 Goldammer, 1997; see also Goldammer and Price, 1997.

103 On zero-burning techniques for oil palm plantation development, see Golden Hope Plantations Berhad, 1997; Wakker, 1998.

104 Blakeney, 1998.

105 Ibid.

106 Prohibiting all burning during severe droughts would undoubtedly visit hardship on traditional shifting cultivators, who would be effectively prevented from planting their crops under such conditions—at least where fire is an integral part of their cultivation system, as is the case in many traditional Indonesian societies. The inequities that such a ban would inflict on traditional shifting cultivators could, however, be ameliorated through the use of food subsidies given in exchange for forgoing burning. This is not a perfect solution, but the Indonesian government, like most other governments, regularly provides a wide range of subsidies to different social groups in order to achieve important societal objectives and could do so in this case during periods of extreme drought.

107 Indonesia already has a fire prevention mascot similar to "Smokey the Bear"— "Pongi the Orangutan." Pongi, however, has not yet been utilized in the intensive and effective way that Smokey has in the United States.

108 GTZ (German Technical Cooperation) Sustainable Forestry Management Project, "Forestry Highlights from the Indonesian Press," April/May 1998.

109 Even if this were a case of police-firm collusion, it still points to a need for procedural legal reforms. The problem would not arise if the provincial environment agency possessed authority to bring a criminal case to court on its own, rather than having to convince the police to do so.

110 GTZ, 1998.

111 Personal communication, GTZ IFFM/SFMP, Samarinda, Indonesia, October 23, 1999.

112 Trenberth and Hoar, 1997.

113 For a review of donor-funded fire projects in Indonesia, see Dennis, 1998.

114 See "Joint Press Statement, Fifth ASEAN Ministerial Meeting on Haze, Malaysia, 30 July, 1998." ASEAN efforts to develop a regional haze action plan were funded in 1999 by a $1.2 million grant from the Asian Development Bank (ADB, "Technical Assistance for Strengthening the Capacity of the Association of Southeast Asian Nations to Prevent and Mitigate Transboundary Atmospheric Pollution." Manila, February 1998, TAR:OTH 32019). A related ADB grant of $1.2 million funded specific activities in Indonesia (ADB, "Technical Assistance to the Republic of Indonesia for Planning for Fire Prevention and Drought Management," Manila, March 1998, TAR: INO 31617).

115 As fires once again spread in Kalimantan and Sumatra in August 1999, Brunei was threatening to sue Indonesia ("Brunei Threatens to Sue Jakarta if Fires Not Contained," Television Corporation of Singapore, July 29, 1999), and WALHI, the Indonesian Forum for the Environment, was calling on Malaysia, Singapore, and Brunei to take Indonesia to the International Court of Justice over its inability to prevent fires and its tolerance of fire-based land clearing by forestry and plantation firms ("Experts Call for Indonesia to Face Court over Smog," *Jakarta Post*, August 6, 1999).

PART II

BEYOND THE FOREST FIRES:

REFORMING INDONESIAN FOREST POLICY

VI. THE POLITICAL ECONOMY OF FORESTS IN THE SUHARTO ERA

A great deal of reporting on the 1997–98 fires in the international media conveyed the impression that vast areas of Indonesia's primary rainforests were going up in smoke. Numerous researchers have pointed out, to the contrary, that most of the burning occurred in secondary and logged-over forests, scrublands, plantations, and agricultural plots.[116] Viewed in a time-bounded perspective, this is true: at the time the fires started, the areas that burned were, for the most part, not primary forest.

But almost all of the areas that burned in 1997–98 had, in fact, been covered by primary rainforest as recently as 30 years ago. (*See Box 9.*) Narrowing the perspective to the situation circa August 1997 leaves us unable to explain the factors that degraded and cleared these vast areas of primary rainforest and reduced them to such a fire-prone state. We are then also unable to explain why so many of the fires were set intentionally and why so little was done to prevent or extinguish them.

> *The forest fires in Indonesia are the visible symptom of structural problems that can only be solved by addressing them at policy, legal, and institutional levels. Clearly spoken: the investment in costly fire emergency operations this year will not stop the fires nor solve the problems. They are only justified in order to save the lives and belongings of people and to protect unique reserves of biodiversity which are global commons.*
>
> Gerhard Dieterle,
> Haze Emergency Coordinator,
> German Technical Cooperation (GTZ),
> Indonesia, April 23, 1998

Since the beginning of the Suharto regime in the late 1960s, a progression of bad policies and practices concerning land and resource allocation and use has brought about rapid forest degradation and the wholesale conversion of many forested areas to either agricultural land or to unproductive and biologically impoverished brushland and grassland. These factors have interacted synergistically to open and destroy the country's forest frontier in

one region after another. Typically, large and poorly managed timber concessions first open the forest and provide access via logging roads. Illegal loggers and small slash-and-burn farmers soon follow, completing the degradation begun by poor harvesting practices. The areas thus degraded are then converted to timber or agricultural plantations or to transmigration resettlement sites. Fire is the cheapest way to clear the remaining vegetation, which is also

much more prone to accidental fires than the intact forest had been. In addition, Indonesian peasants have long used fire as a defensive weapon against the takeover of their lands by outsiders, and fire has been used as a weapon of conquest by commercial interests seeking to take over forestlands from local communities.[117]

In short, the fires of 1997–98 were the logical and inevitable result of long-standing struggles over the control of forestlands and resources and a reflection of the imbalances and abuses of power that characterized New Order natural resource policies for three decades. The result is a situation in which vast areas of degraded, fire-prone forestlands already exist and processes are at work that are increasing the amount of such degraded land.

---

9 THE FORESTS OF KALIMANTAN AND SUMATRA BEFORE THE SUHARTO ERA

Until the mid-20th Century, Sumatra and Kalimantan were for the most part forest-covered. They were only sparsely populated by forest dwellers who subsisted by hunting, practicing swidden agriculture, and gathering food and natural products from the forests. Since some of those forest products were in demand in Java, other parts of Asia, and later Europe, trade networks developed on major river systems, usually under the control of a ruler based at a trading center at the river's mouth. Forest-dwelling communities established use rights to large areas for hunting and collecting.

Around the beginning of the 20th

Century, the Dutch colonial administration began a long-running debate on the right of forest-dwelling communities to harvest and sell nontimber forest products versus the right of the state to control and tax these products.[1] At that time, however, timber from the interior forests was not valuable enough to justify the cost of harvesting it and transporting it to market. Several firms attempted to log the most accessible forests during the later colonial period, but only a few relatively small timber concessions were commercially viable.

Aside from the establishment of tobacco and rubber plantations in eastern Sumatra, oil and coal extraction in East

and South Kalimantan, and limited timber harvesting in Kalimantan, the forests remained unaffected by commercial exploitation until the late 1960s. This is not to say that the forests were untouched or that the two islands shared the same land-use history; land-use practices and population densities of forest dwellers developed differently on Kalimantan and on Sumatra. In general, Sumatra led Kalimantan in growth of rural population density, intensification of agriculture, and orientation toward market crops. Within Kalimantan, West and South Kalimantan developed more rapidly than Central and East Kalimantan.

During the violent years of World War II and the subsequent turbulence of the independence movement and the early postindependence period, the forests of the outer islands were not logged, due to political and economic instability. After World War II, rainforests began to be harvested elsewhere in Southeast Asia, and demand grew for timber species of the dipterocarp family. The chain saw, together with modern harvesting and road-building equipment, made logging tropical rainforests technically feasible and profitable.[2]

Notes:
1. Potter, 1988.
2. Whitmore, 1984.

The four most important factors in the recent and ongoing degradation of Indonesia's forests are:

- the logging industry, which since the early 1970s has laid claim to nearly two-thirds of the nation's land area;
- the government's push, since 1990, toward rapid development of industrial timber plantations to supply raw materials for the growing pulp and paper industry;
- the rapid development of oil palm plantations;
- the government's transmigration program, which resettles people from densely populated Java onto the forest frontiers of the country's larger, less populous islands.

As a result of these ill-advised land-use policies, East Kalimantan has since 1985 lost the single greatest amount of forest of any province. In 1985, satellite mapping revealed that some 90 percent of the province was still forested. By 1998, that figure was down to 68.5 percent, representing a loss of at least 4.5 million ha.[118] A survey during 1998 and 1999 concluded that the 1997–98 fires by themselves affected 5.2 million ha (see Map 4), so the total area of forest lost between 1985–2000 may well be significantly higher.[119]

All four of these factors are animated by the same cross-cutting characteristics of Suharto-era natural resources policy:

- centralized, top-down decision-making processes made by sectoral bureaucracies strongly biased toward the interests of a small group of businessmen with close ties to the Suharto family and other members of the ruling circle;
- lack of effective legal or administrative mechanisms to hold bureaucrats and their corporate clients accountable for violations of the law, usurpation of local access to resources, corrupt practices, and other abuses of power;
- a systematic abdication of the central government's role in monitoring use of natural resources, in favor of private sector "crony capitalists." These interests were granted large natural resource exploitation concessions, often in collaboration with provincial governors and other local officials intent on promoting rapid economic growth (and their own enrichment);
- policies and practices that have steadily eroded the legitimacy and function of customary (adat) rights and management systems related to natural resources, and have thus deprived forest-dependent communities of their long-standing access to forest resources.

LOGGING POLICIES AND PRACTICES IN THE SUHARTO ERA

When the Suharto government came to power in the mid-1960s, economic planners took immediate steps to develop Indonesia's weak economy and began to develop the legal framework to permit private firms to harvest and export timber. Sumatra and Kalimantan were the first targets of forest exploitation because they had the largest stocks of commercially valuable tree species and were closest to Asian markets.

The Forestry Act of 1967 provided the legal basis for awarding timber harvesting rights, and many large 20-year concessions were granted soon afterward. Exports of unprocessed logs rose dramatically in the 1970s, providing foreign exchange, capital to build Indonesia's emerging business empires, and employment. From 1969 to 1974, for example, nearly 11 million ha of logging concessions were granted in East Kalimantan alone.[120] Whereas in 1967 only 4 million m³ of logs were cut from Indonesian forests—mostly for domestic use—by 1977 the total had risen to approximately 28 million m³, at least 75 percent of which was for export.[121] Gross foreign exchange earnings from the forest sector rose from $6 million in 1966 to more than $564 million in 1974. By 1979, Indonesia was the world's major tropical log producer, with a 41 percent share ($2.1 billion) of the global market, representing a greater export volume of tropical hardwoods than all of Africa and Latin America combined.[122]

Roads, towns, and other infrastructure were built in Sumatra and Kalimantan in the wake of the timber bonanza, and the populations of these islands grew substantially. The population of East Kalimantan, where a simultaneous oil boom was occurring, doubled between 1970 and 1980, transforming the landscape as agricultural settlers followed the loggers into the forests.[123]

The timber industry went through a period of consolidation in the early 1980s when a ban on log exports was imposed, creating a few enormous vertically integrated timber firms that concentrated on plywood production. The number of plywood mills in the country grew from 21 in 1979 to 101 in 1985. Plywood production rose from 624,000 m³ in 1979 to nearly 4.9 million m³ in 1985 and to more than 10 million m³ in 1993, nearly 90 percent of which was exported. At the same time, the industry became increasingly concentrated in the hands of a small number of firms with connections to the regime. By 1994, the top 10 groups controlled nearly 24 million ha (37 percent) of the 64 million ha of logging concessions in the country; the share was 64 percent in timber-rich East Kalimantan. These big firms formed a cartel that made Indonesia the world's largest plywood producer and succeeded in raising international plywood prices.[124]

By mid-1998, more than 69 million ha of forest area had been allocated to 651 concessions. Of that area, 49 percent was operated by concessionaires who were in their first 20-year term and 22 percent by concessionaires whose term had been renewed. The remaining 29 percent (21.2 million ha) remains mostly in limbo: 9.5 million ha are slated for "rehabilitation," 8.3 million ha are reserved for as yet unallocated "other uses" (timber and oil palm plantations and transmigration sites), and 3.3 million ha are to be restructured as joint private sector-state forestry corporation concessions. Overall, forest degradation from logging concessions totaled some 16.6 million ha by mid-1998.[125]

The most basic problem with the government's management of logging is that land was designated as production forest with little knowledge of the characteristics of the land, the traditional rights of communities already living there, or the conservation importance of forest ecosystems. The negative effects of uninformed land allocation decisions were exacerbated when concessions were awarded to companies and individuals with no experience in timber harvesting, supervised by forestry officials who lacked the political support, incentives, and resources to provide meaningful oversight of harvesting operations.[126]

By mid-1998, more than 69 million ha of forest area had been allocated to 651 concessions.

These weak forest management institutions have resulted in inefficient extraction of timber, unnecessary damage to the remaining trees, excessive waste wood left in the forest, unnecessarily severe impacts on animal populations, soil erosion, and stream pollution.[127] Low government royalties on timber and weak performance supervision give the concessionaires little incentive to reduce timber waste, mitigate environmental impacts, or manage their concessions sustainably.[128] Virtually all of the lowland forest in Sumatra and most of the economically harvestable lowland forest in Kalimantan has been logged, leaving behind a legacy of social and ecological disruption, with little thought to managing the logged forests sustainably.

The state-sponsored expansion of the plywood industry in the 1980s created considerable overcapacity in relation to the amount of timber Indonesia's forests can sustainably produce. In September 1998, the minister of forestry and plantations predicted that the wood-processing industry would face an annual log shortage of at least 25 million m³ over the next five years. As of mid-1998, that industry officially included 1,701 sawmill companies, with a combined annual production of 13.3 million m³, 115 plywood firms with installed capacity of 8.1 million m³, and 6 pulp and paper companies with production capacity of 3.9 million m³. Taken together, and producing at full capacity, these industries need 57 million m³ of timber, while the officially designated annual cut for the next five years is set at 31.4 million m³.[129] This cutting target is in fact much higher than other estimates of a sustainable cut. The World Bank's 1993 Indonesia Forestry Sector Review, for example, argued that a realistic level would be only 22 million m³ per year.[130]

The most basic problem with the government's management of logging is that land was designated as production forest with little knowledge of the characteristics of the land, the traditional rights of communities already living there, or the conservation importance of forest ecosystems.

In any case, the rate of cutting is widely thought to be much higher than that officially reported. The World Bank sector review reported that for every cubic meter cut, at least an equal amount of usable wood is left behind and that at least 8 million m³ are left rotting in the forest every year.[131] In addition, illegal logging is widespread and systematic in many parts of Indonesia. Illegal removals are thought to be in the range of 30 million m³ per year, exceeding legal cutting.[132] Illegal timber brokers flourish throughout the country, supplying processors who cannot obtain adequate supplies legally.[133] Logging concession roads often provide illegal loggers with access to the forest, encouraged by the lack of meaningful access controls by either the logging firms or local forestry officials. Widespread and systematic illegal logging in two of Indonesia's showplace national parks, carried out in collusion with local authorities, was extensively documented by a research team and reported in the media during 1999.[134] The two parks, Gunung Leuser in northern Sumatra and Tanjung Puting in southern Kalimantan, are the two most important protected habitats for the orangutan in Indonesia.

A 1998 analysis by the Center for International Forestry Research (CIFOR) predicted that the economic crisis is likely to intensify forest degradation from logging.[135] In early 1998, demand for Indonesian plywood from its major markets in East Asia had collapsed, and the international price of plywood had plunged from $500 per m³ in 1997 to $300 per m³. But, in April, demand for Indonesian plywood in China and other Asian countries surged, while Malaysia, another important supplier, was curtailing its exports. Prior to the crisis and the forest fires, the plywood sector was already facing severe supply difficulties due to overharvesting and poor logging practices. The forest fires destroyed significant amounts of the timber stock, and, as noted above, a significant supply shortfall for wood processors is predicted for the next five years. "The combined effect of the low price for Indonesian plywood and potentially high demand, and restricted supply resulting from the fires means that producers will search for stems in ever more remote and inappropriate places. The potential for increased damage in production forests and unauthorized logging in recently logged production forests and some protection forests appears to be high."[136] There are also indications that the rising price of kerosene is causing many people to turn to wood fuel.

In short, logging has been a major factor in degrading Indonesia's forests and will continue to be so unless fundamental policy reforms are enacted and implemented. But logging has been only the first stage in the process of deforestation.

INDUSTRIAL TIMBER PLANTATIONS

About the time of the Fourth Five-Year Development Plan (1984–1989), the Indonesian government launched an ambitious plan to establish vast areas of monocultural fast-growing timber plantations, particularly in Sumatra and Kalimantan. The plan was accelerated around 1990. At the outset the government justified the program as a way to augment supplies of timber from natural forests and promote nature conservation.[137] To this ostensible end, timber plantation entrepreneurs have received interest-free loans from the "Reforestation Fund" collected from logging concessions. In addition, under a joint program of the Ministries of Forestry and Transmigration, introduced in 1992, the government can supply 40 percent of investment, plus labor from specially established transmigration settlements, while investors supply the remaining capital. By the end of 1994, almost 39 percent of the area planted was in transmigration estates.[138]

The timber estate program got off to a slow start. In the late 1980s, the government was planning to open 1.5 million ha annually and to reach a total of between 4.4 million ha and 6 million ha by 2000. By 1998, 2.4 million ha had been established.[139] (*See Table 4.*) As of May 1998, the government had approved applications for 4.6 million ha in timber and pulp plantations, nearly 70 percent of which would be for pulp production.[140]

Despite its professed intentions, the timber estate program has in fact become a powerful engine of defor-estation and is currently almost totally devoted to providing feedstock for the rapidly growing pulp and paper industry, which is annually adding some 13 million m³ of demand that would not exist without this industry.[141] Plantations have often been estab-lished on degraded timber concessions by the very same firms whose poor logging practices degraded the forest in the first place. As the World Bank points out, "logging operations can degrade a site with little risk of serious penalty, and in the process set them-selves up to receive a license to convert the site so damaged into a HTI [timber plantation] or tree crop estate."[142]

> *Despite its professed intentions, the timber estate program has in fact become a powerful engine of deforestation.*

Indonesia's ambitious plan to become a major pulp and paper producer is thus multiplying indus-trial demands on the forest resource base. Pulp production rose from 1.1 million tons in 1991 to 3.1 million tons in 1996, with pulp plantation projects covering some 5.1 million ha.[143] An added 10 million tons of new pulp capacity is planned by 2005, according to the executive director of the Indonesian Pulp and Paper Association, although that target is unlikely to be met in the current climate of economic crisis.[144]

Feedstock for this intensive program will eventually come from short-rotation plantations on already degraded forestlands. By 1993, 33 potential pulpwood plantation concessions of 200,000 ha to 300,000 ha each had been identified, although as of June 1997 the government had allocated only 2.63 million ha to 13 firms.[145] In reality, only 60,000 ha to 80,000 ha of each concession are actually being planted with new trees. The remainder of these plantations, usually logged-over but sometimes unlogged primary forest, are cut to supply the designated mill operation until the rotation planting can supply pulpwood.[146] Demand for pulp feed-stock thus competes with timber demand in firms' investment deci-sions on plantations (pulp versus timber species). If all planned pulp and paper mills actually come on stream, as much as 30 million m³ of natural forest will have been used for pulpwood by the end of 2000.[147]

	ALLOCATED	REALIZED BY 1998
Sumatra	2,148,964	893,463
Kalimantan	2,928,414	956,261
Sulawesi	255,791	85,455
Maluku	64,775	77,656
Irian Jaya	153,250	39,996
Other	48,730	352,215
Indonesia	5,599,924	2,404,364

TABLE 4

Timber Plantation Development to 1998 (hectares)

Source: World Bank, 1999c.

The rapid expansion of pulp-wood and other timber plantations has led to numerous conflicts with local communities. Although the effects of logging concessions on local communities can be onerous, people are still able to retain some access to forest resources in the concessions. Plantations, however, and the clear-cutting that accompanies them, impose a much higher level of deprivation on communities that depend on the forest areas in question for their livelihoods.[148]

As noted above, industrial timber plantation firms were blamed for intentionally setting fires to clear land in 1997. Of the 176 plantation firms identified as culprits by the Forestry and Environment Ministries in September 1997, 28 (16 percent) were industrial timber plantations.[149] In light of the Environment Ministry's statement that "it has been proven that 85 percent of the fires were set by oil palm and industrial timber plantation firms,"[150] one can conclude on the basis of that estimate that approximately 14 percent of the fires were set to clear land for timber plantations.

THE OIL PALM BOOM

Palm oil, extracted from the fruit of a species of palm originating in Africa (*Elaeis guineensis*), is widely used as cooking oil and as an ingredient in soap, margarine, and a variety of other products. Global production grew from 14.7 million tons in 1994 to nearly 16 million tons in 1997. Production in that year was dominated by Malaysia, the largest producer (with 50.6 percent), and Indonesia (28.8 percent), the second largest. Global production is expected to grow by more than 7 percent annually for the foreseeable future, and by 2005 Indonesia is expected to produce some 12.2 million tons, or 41.4 percent of the total.[151]

Expansion of oil palm plantations is probably the largest single commercial force behind deforestation in Indonesia; the area covered by these plantations grew from about 843,000 ha in the mid-1980s to nearly 3 million ha in 1998.[152] (*See Table 5.*) Of this area, 46 percent was held by private companies, with smallholders and older state-run plantations making up the rest.[153] Most plantations are currently in Sumatra, but Kalimantan is being rapidly developed, and Irian Jaya is the primary target for future expansion. According to a recent study, "it can be said that almost all of the existing oil palm plantation areas result from the conversion of production forest." This is because the procedure for acquiring forest-land is relatively easy and the firm can clear-cut and sell standing timber, a profitable side business. As of 1997, the agreed area of production forest to be converted for plantations had reached 6.7 million ha, in addition to 9 million ha proposed for further development of tree crop plantations on other lands.[154]

Expansion of oil palm plantations is probably the largest single commercial force behind deforestation in Indonesia.

The Suharto government aimed to reach a total of 5.5 million ha of oil palm plantations by 2000—a target that was not met. Three million ha have been established, and an annual conversion rate of 200,000 ha to 250,000 ha per year seems likely.[155]

Indonesia's oil palm industry is dominated by some of the same domestic conglomerates that control the logging, wood-processing, and pulp and paper industries. Just four companies held 68 percent of the 1 million ha of estates in private hands in 1997.[156] There is also considerable foreign investment: as of the end of 1998, 50 foreign firms were involved in the oil palm sector, with total investments valued at $3 billion.[157] As with timber plantations, the rapid expansion of oil palm plantations has given rise to widespread conflicts with local communities.

TABLE 5

Oil Palm Plantation Development in Indonesia, mid-1980s to 1998 (hectares)

	Oil-Palm Area, Mid-1980s	Oil Palm Area, 1998	New Oil Palm Area Since Mid-1980s	Outstanding Applications From Developers, 1995
Sumatra	805,800	2,240,495	1,434,695	9,395,697
Kalimantan	0	562,751	562,751	4,760,127
Sulawesi	11,800	101,251	89,451	665,379
Maluku	0	0	0	236,314
Irian Jaya	23,300	31,080	7,780	590,992
Other	1,800	21,502	19,702	1,777
TOTAL	**842,700**	**2,957,079**	**2,114,379**	**15,650,286**

Source: World Bank, 1999c.

Indonesia's haste to expand the industry, and the privileged political position of the major firms, has made land clearing for oil palm the largest single forest fire risk factor in Sumatra and Kalimantan. Former minister of agriculture Baharsja estimated that 550,000 ha could have been cleared by burning in 1997, as this was the amount of land targeted for conversion to pulp and palm oil plantations and for agricultural land settlement schemes.[158] Burning is attractive to plantation firms because it removes waste wood and vegetation rapidly and requires relatively little heavy equipment or technical expertise.[159]

The minister of forestry at the time, Djamaluddin Suryohadikusuma, announced that 46 percent of the hot spots appearing on satellite images on September 28, 1997, were in lands granted for plantations.[160] Ironically, plantation firms suffered heavy losses later in 1997 and in 1998 as fires spread out of control into established plantations. In early April 1998, the minister of state for environment, Juwono Sudarsono, estimated that 160,000 ha of plantations had been damaged in East Kalimantan during the previous three months.[161] A researcher investigating the role of plantations in the fires found that it was not uncommon for local people to purposely set fire to tree crops to protest loss of their land to plantation firms.[162] The same researcher also interviewed farmers who believed that plantation firms deliberately set fire to their crops to reduce the compensation owed the farmers for being displaced by plantations.

THE TRANSMIGRATION PROGRAM

Between 1969 and 1993, transmigration—the government's program for resettling people from densely populated Java and Bali to Sumatra, Kalimantan, and the other "outer islands"—opened 1.7 million ha of agricultural land and transported some 8 million people.[163] The program affects a much greater area, however, due to poor site choices and the land-clearing practices employed. A 1994 World Bank evaluation of the $560 million in loans it made to Indonesia for the program during the 1970s and 1980s concluded that land clearing was not carried out according to agreed legal guidelines. Slopes of over 8 percent had been cleared, trees had been bulldozed into waterways, anti-erosion measures along contours had not been taken, and no attempt had been made to harvest the commercial timber left partly burned in the field. The effects on local communities, particularly traditional indigenous groups, have been extremely negative. In the case of the forest-dwelling Kubu of Sumatra, for example, the report concluded that "there has been a major negative and probably irreversible impact."[164]

Over the past decade, the emphasis of the transmigration program has shifted away from subsistence agriculture and toward wage labor on industrial timber estates and oil palm plantations. As noted above, almost 39 percent of the timber estate area planted lies in transmigration sites,[165] and some 956,257 ha of oil palm plantations with a formal link to transmigration sites had been established by the end of 1995.[166]

THE MILLION-HECTARE PEAT-SWAMP PROJECT IN CENTRAL KALIMANTAN

Beginning in 1995, the Suharto regime embarked on its last and most disastrous megadevelopment project—a scheme to transform peat forests covering more than 1 million ha in the heart of Kalimantan into a rice-growing region colonized by more than 1.5 million transmigrants from Java. The region in question constitutes a large part of the largest peat-swamp floodplain in western Indonesia and contains some of the oldest and deepest peat deposits on the planet. It is home to the largest contiguous population of orangutan in the world, as well as countless other rare and endangered species of flora and fauna. Thousands of indigenous Dayak people have lived in the region for centuries, benefiting from the rich fish harvests in the swamps and rivers and harvesting numerous nontimber forest products such as rattan.

Peat swamps are complex and fragile ecosystems that are essentially unsuitable for large-scale agriculture because of their hydrology, which is "very difficult to manage," and their extremely acidic soils on which "it

is impossible to grow economically viable crops." In addition, their high rates of subsidence when the area is drained for crops causes "considerable difficulty in the establishment of successful plantation agriculture" particularly for "top-heavy" crops like oil and coconut palm.[167]

This was, nevertheless, the place that Suharto decided was suitable for massive, intensive settlement and agricultural development. Such was the nature of his regime that not one of his ministers dared point out that the project was doomed to fail—a conclusion that virtually anyone would reach after an hour's reading in the extensive literature on development in tropical peat swamps.[168]

The project, commonly known by its Indonesian acronym PLG, perfectly illustrates the close linkages between New Order forest and land-use policies and the fire disaster of 1997-98. The regime's ill-advised policies on forestry, plantations, transmigration, and land claims by indigenous people all came together in this debacle, causing untold environmental damage, extensive human suffering, and one of the worst single concentrations of fires in 1997, with massive releases of carbon into the atmosphere.

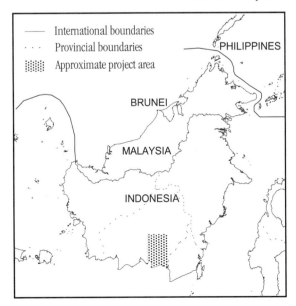

The PLG project was promoted by Suharto himself and was made state policy in 1995 by Presidential Decrees No. 82 and 83. Situated in Central Kalimantan province, the area intended for the project covered about 1.4 million ha. Of this, some 500,000 ha was to be converted from peat-swamp forest to rice and other crop cultivation (adding to the existing 80,000 ha of rice land in the project area), with the remaining areas left under forest cover. More than 700 km of large canals, 25 m wide and 5–6 m deep, as well as numerous secondary and tertiary canals, were to be constructed to supply irrigation water and transport. (*See Maps 5a and 5b.*) Some 672,000 transmigrants from Java were slated to be settled in the area, nearly 400,000 of them by the end of 1999,[169] with the total eventually rising to 1.75 million (350,000 families).[170]

Forest clearing and construction of canals began in late 1995, six months before the required environmental impact analysis (EIA) was initiated. The EIA concluded that only about 30 percent of the project area was at all suitable for agriculture—a finding ignored by the government and its contractors.[171] By 1997, some 13,500 transmigrants had been brought in, and the project area had become a free-for-all zone for illegal logging facilitated by deals between the companies building the project infrastructure (canals, transmigration sites, and roads) and small logging companies. Among the contractors was PT. Rante Mario, owned by the president's youngest son, and subsidiaries of the Salim Group, controlled by a long-time Suharto business partner, Liem Soe Liong.[172] By October 1997, floats of hundreds of logs of the area's best timber were being continuously sent downriver to the coast from the project area.

"How could a grossly stupid project of this magnitude happen in this modern age of advanced agronomic technology and environmental awareness in which sustainable development is the watchword? The answer lies with the autocratic government of President Suharto..."

Jack Rieley, Director
Kalimantan Tropical Peat Swamp Forest Research Project
School of Geography, University of Nottingham,
United Kingdom
(Rieley, 1999)

Meanwhile, dozens of portable sawmills worked constantly along the riverbanks to turn poorer-quality wood—that would rot in a few years—into housing for transmigrants.[173]

Fire was used systematically and pervasively to clear project lands. (*See Map 5c.*) Hundreds of temporary migrants moved up the rivers and canals into the area, seeking work with the construction companies or the illegal loggers and sawmills. Many were hired by the contractors to set fires. One project foreman estimated that about 90 percent of the workers moving around his canal construction site were not officially working for the project.

By mid-1997, the combination of drought, large amounts of logging waste, and intentional burning had created an inferno that raged across huge areas of the million-hectare project site from July through November. Because of the area's high concentrations of peat swamp,

these fires were particularly smoky and largely accounted for the intense haze that blanketed Malaysia's Sarawak state during September. If the project was not the single largest source of haze, it was certainly one of the largest. Yet the government did not officially acknowledge the key role of the president's pet project in polluting much of Southeast Asia.[174]

A journalist reporting from the PLG area in October 1997 accurately described the devastation:

"Suharto's grand project is today one of the most desolate spots on earth–a vast, stinking, blackened, smouldering and toasted place. Thousands of square kilometers of land are shrouded in smog, as the earth itself burns from deep below in the peat of would-be paddy fields. ... The whole hydrology of the area has been affected, with the water table dropping several feet. The river has turned a bright green and is

"Frodo and Sam gazed out in mingled loathing and wonder on this hateful land. All seemed ruinous and dead, a desert burned and choked. . . . Smoke trailed on the ground and lurked in its hollows, and fumes leaked from fissures in the earth."

J.R.R. Tolkien,
The Lord of the Rings

mostly undrinkable. And everywhere the land has been systematically and deliberately torched." [175]

Meanwhile, the project had failed spectacularly as a rice-growing effort. Suharto's 1995 decree had specified that rice had to be harvested from the project within two years, and about $350 million had been spent to that end (much of it, ironically, taken from the Reforestation Fund).[176] "That is an order that must be obeyed," observed peat swamp expert Tejoyuwono of Gadjah Mada University. "[Officials charged with implementing the project] have just been making it up as they go along. They have ignored all established technical procedures."[177] Because the soils are unfit for agriculture, transmigration site officials confided to one of the authors that they had to use up to 6 tons of lime per hectare to improve the pH of the soil enough to support crops, which still were not growing very well. Rieley (1999) reported that by mid-1999, "not one fruiting head of productive rice has been grown and a landscape of one million hectares (the size of Northern Ireland) lies devastated and useless."

Lack of water was another major problem; the poorly planned canals had actually cut off water supplies to many planned agricultural areas. Ignoring the information of his own subordinates—the district chief had told the press two days before that "most of the irrigation canals in the area have gone dry"[178] —the governor of Central Kalimantan in early September asked, "Who says there is a lack of water? The agriculture project director there has already fixed the problem."[179]

Of the planned 2,500 ha of rice that were to be harvested in mid-October—with Suharto in attendance—only 279 ha were growing by the end of August.[180] After much suspense in the local press, the president did not show up for the harvest after all; the "thick smoke" blanketing the region made travel too difficult, it was explained.[181] Indeed, by October 1, about 5,000 ha in the area immediately around the two already inhabited transmigration areas were aflame. Local residents reported to the press that the fires had been burning since early August but that they had observed "no efforts whatsoever" to put out the fires by either the government or the project contractors. The PLG project director, however, told the press that this fire was in the "small category" and hence not much of a problem.[182]

Difficult as life has been in the PLG transmigration areas, it has been easy compared with the suffering of the indigenous Dayak people in the area. The Dayaks in the seven villages along the Mengkatip River—the first sector to be developed within the project area—have seen their lives largely destroyed by the project and by the fires that it caused. Until 1996, these people had lived a relatively prosperous life practicing traditional agriculture, cultivating and selling large quantities of rattan from carefully tended forest gardens, and exploiting the bountiful fish stocks found in peat-swamp pools (beje). When construction of the PLG canals and the associated land clearing began in April 1996, the first result was that the beje dried up and the fish died. The massive runoff of lime from the transmigration sites turned the Mengkatip River bright green, and, according to local fishermen, all but two species of the formerly plentiful river fish stocks completely vanished.

> *One million hectares, one million wounds.*
>
> Graffiti in Dadahup village,
> Central Kalimantan

Traditionally, the Dayak have claimed an area 5 km back from both sides of the river as their traditional (*adat*) land, using it for intensive rattan cultivation, farming, fishing, hunting, and collection of grasses and other useful products. With the arrival of the PLG project, fully 50 percent of their land was confiscated, and many areas of rattan and other crops were destroyed by the construction of canals. Then, in mid-1997, the fires set to clear land for the project spread out of control and destroyed most of the rattan areas in the narrow strip along the river that the government had spared.

Local village leaders conducted a systematic inventory of their losses. Together with the Indonesian Forum for the Environment (WALHI), they made a conservative estimate—since many people's losses could not be included in the survey—of Rp 20,910,533,000, or nearly $7 million at mid-1997 exchange rates.[183] (*See Table 6.*)

When aggrieved local leaders sent a delegation to the district head (Bupati) to demand compensation for at least some of their losses, he told them (and the press) that "in line with policy from the Center concerning this national project, there will be no compensation whatsoever paid," since the project was for the good of the people anyway. He noted, however, that the complainants would be given the chance to become "local transmigrants" in the nearby transmigration site—the one where agriculture had almost totally failed and houses were being built with substandard timber.[184] This pronouncement flatly contradicted a formal April 1996 agreement, signed by the heads of the provincial and district development planning agencies, to pay compensation for standing rattan and other crops at stated rates.[185]

By mid-1998, the economic crisis and the fall of Suharto had placed the eventual fate of the PLG project in question. In May 1998, at the request of the minister of public works, a World Bank water resources engineer conducted a short survey of the PLG project area. His report described in detail the scope of the disaster caused by the project. Noting that almost all of the primary canals were aligned over deep and medium-depth peat conservation areas, the engineer warned that subsidence of the peat would occur rapidly near the canals, compromising their drainage functions and causing increasing susceptibility to fire. He also noted that no information was available on damage to wildlife habitat or about the impacts on the local indigenous people. He concluded by calling for a halt to further expansion of the project and a complete reevaluation, including a new EIA. Other sobering conclusions included the following:

TABLE 6

Losses Incurred by Seven Villages in the Mengkatip Watershed Resulting from Land Appropriation and Fires Associated with the Million-Hectare Rice Project

NATURAL RESOURCE	AMOUNT	ANNUAL PRODUCTION	FINANCIAL LOSS (RUPIAH)
Rattan burned	3,492 ha	17,463 tons	4,017,375,000
Rattan confiscated	4,070 ha	20,351 tons	4,680,667,000
Fruit tree orchards		226,870 m3	6,816,111,000
Fishponds confiscated	487 tons	487 tons	730,500,000
Fishponds burned or dried up	1,200 tons	1,200 tons	1,800,000,000
Purun grass	75 stands	432 bundles/ ha	1,701,000,000
Rice fields	382 ha	764 tons	1,146,000,000
Rubber	504 ha	5 tons	18,900,000
TOTAL			**Rp 20,910,553,000**
			US$6,970,184[a]

Note: a. Calculated on a rate of Rp 3,000 = $1, the exchange rate in mid-1997.
Source: WALHI, 1999.

"The whole water system needs to be redesigned and made compatible with soil and topographic constraints. It is not unlikely that—to reverse the damage done—the major canals will have to be filled in . . .

At least Rp1.5 trillion ($500 million) has been spent on the project to date. Redesign of the project means an even greater expenditure than ever foreseen. At the same time the agricultural area will be reduced to below 500,000 ha while rice yields and agricultural economics are uncertain. No benefit-cost engineering-economic and sensitivity analysis has been undertaken to date, especially one that includes settlement and environmental mitigation costs as well as other irreversible losses. GOI [Government of Indonesia] owes itself such an analysis in order to determine the desirability of further major expenditure on this project."[186]

Further analysis of the project's canal system has revealed that rather than irrigating the peat areas, the canals have served to systematically drain their moisture into the sea because the land's topography was not taken into account.[187] As a result, the water table is falling, the remaining vegetation is dying off, and the peat is shrinking by 1 to 2 centimeters per year—releasing large amounts of carbon and increasing fire risk as the land dries out. Poor design, construction and maintenance have also resulted in rapid silting-up of the canals, and many will be filled in with peat within five years. In the words of another expert:

"The channels have not, and never could function, owing to the contours of the land surface and the constraining physical properties of the peat itself. After only two years the main channels are losing their water, their banks are collapsing and they are silting up with peat mud. They are already in a state of disrepair but are still being used as conduits along which people gain access into the interior. As a result all remaining timber is being removed and, in the process, debris is set alight and the surface peat catches fire generating the dense unhealthy haze that has beset Southeast Asia in recent years."[188]

In June 1998, the minister for transmigration and resettlement of forest encroachers (the actual name of the ministry) announced that the government would resurvey the whole project area before making a decision about the project. He acknowledged the numerous calls to halt the project (based largely on the findings of the World Bank survey) but argued that "if we stop the project just like that, it would be the same as throwing away Rp1.4 trillion."[189]

In August 1998, over 120 representatives from communities whose livelihoods had been destroyed by the project (including many from the Mengkatip River area) occupied local government offices in the Central Kalimantan capital of Palangkaraya for a full week. They demanded that the project be halted, all lands returned to their rightful owners, damaged ecosystems restored, and compensation paid for local financial losses. Local government officials replied that the project was run by the central, not the local, government and so was not their affair.[190]

The following month, the minister for transmigration and resettlement of forest encroachers confirmed to the press that reevaluation of the project was under way. But rather than announcing the ecological restoration and careful assessment favored by the World Bank's evaluator, or the complete halt and payment of reparations demanded by the local people, he stated that the areas originally slated for rice and other food crops would be offered to foreign investors to develop oil palm plantations. "Right now, we already have one Japanese investor who is interested. . . . We have to admit that this step will be difficult, because our NGOs are fierce. But according to me, exploiting the peat forests [for oil palm] constitutes nature conservation."[191]

As criticism of the project mounted—and the government ran out of funds due to the economic crisis—then-President Habibie finally issued a decree (No. 80/1999) in July 1999 that formally recognized the failure of PLG, revoked Suharto's 1995 decrees, and essentially ended the project in its present form. Future development of the devastated PLG area has been incorporated into development of a surrounding 2.8 million ha "economic development zone" established by Presidential Decree No. 170 in 1998. Like PLG, the overriding premise in this new strategy is that of land conversion to food crops and plantations, especially oil palm and rubber. There is little reference to environmental protection, and what is mentioned, according to Rieley (1999), is "inappropriate, insufficient and of low priority." No

mention has been made of rehabilitating the nearly one million hectares that is a "wasteland with little prospect for either economic development or hydrological and wildlife conservation."[192] And no plans have been announced to compensate local communities for the immense losses they have suffered.

In mid-1999, however, the Indonesian Forum for Environment (WALHI) brought a lawsuit against the government, including the president and eight cabinet ministers, seeking damages for the destruction caused by the PLG project. The suit alleges that the project misappropriated Rp 527.2 billion (US$65.9 million) from the Reforestation Fund, while ignoring an environmental impact analysis and the aspirations of local indigenous communities.[193] At this writing the case is in progress.

The wasteland left in the wake of the PLG project has set the stage for another massive conflagration in the next long dry season. Indeed, fires broke out again in the PLG area in August 1999; the media blamed farmers who were burning waste from earlier land clearance to get rid of rats threatening their rice harvests.[194] It seems more likely, however, that this renewed burning was at least in part a continuation of the cycle of illegal logging, burning, and land clearing by commercial interests that has characterized the area since 1996.

MEGAPROJECT MADNESS: THE MAMBERAMO BASIN PROJECT IN IRIAN JAYA

The Mamberamo basin is a huge river system of more than 7.7 million ha stretching from the central mountains of Irian Jaya across lowlands and marshes to the north coast. Most of the area is still covered with tropical rainforest, although two-thirds has been allocated to timber or plantation concessions. Its unique wildlife includes crocodiles, tree kangaroos, cassowaries, parrots, and birds of paradise. According to Indonesian government figures, the area has about 7,000 inhabitants, mostly indigenous communities living a seminomadic life of hunting, fishing, practicing horticulture, and harvesting sago palms.

In an echo of the Central Kalimantan rice project (PLG) debacle, the Indonesian government has set in motion plans to carve up Mamberamo for heavy industry, smelting, plantations, rice cultivation, and logging. Dreamed up by former President B.J. Habibie when he was minister of science and technology, this megaproject is expected to take 20 years to complete. The terms of reference for the project published in 1996 show that the scheme, like the PLG project, promises to be a logistic nightmare, with 10 government departments and bodies named as executing agencies and with 12 subprojects.

The project centers on harnessing the power of the 650-km Mamberamo

River through a series of large dams to produce electricity that will transform the region. Upstream areas will be used for dams, agroindustry, and logging, while industrial estates, new settlements, and transport and other infrastructure will be created downstream. Plans include a steelworks, metal smelters, a pulp and paper factory, and a petrochemicals plant, forming the biggest industrial complex in eastern Indonesia.

As in the PLG disaster, the project area is being touted by its promoters as a future food supply center of national importance, with possibly 1 million ha, to be irrigated from the dam scheme, set aside for rice cultivation. And as in the PLG case, there are plans to resettle transmigrants—about 300,000 families, or 1.5 million people—from the western parts of Indonesia to provide the workforce for agricultural projects.

The idea of building large dams and industrial complexes in such a geologically unstable zone is questionable. In February 1996, the island of Biak off the north coast of Irian Jaya was hit by an earthquake measuring 7.0 on the Richter scale. Another, registering 4.6, struck the Mamberamo area in September 1997.

Will the project actually happen? As of mid-1998 the government had already held two workshops for potential

Indonesian and foreign investors, and preliminary studies are now in progress. Reports from the region state that the process of land appropriation has already begun, with the authorities using bribery, threats, and trickery to take land from local people. Some foreign investors have reportedly expressed interest. In April 1997, a government workshop on the megaproject was attended by private companies from France, Germany, Japan, and the Netherlands, as well as Indonesia. In February 1998, Barnabas Suebu, a former governor of Irian Jaya, announced that Australia, Germany, and Japan had agreed to invest in the project, and German and, to a lesser extent, Australian funding has reportedly been used to support many of the feasibility studies.

Despite a bad drought, Irian Jaya did not experience fires during 1997–98 on the same scale as Kalimantan and Sumatra, in large part because much of the territory is still forested and the reckless land conversion so common on the other islands is still relatively limited.

The Mamberamo project is so ambitious and ill considered that it is unlikely ever to be completed. An official of the government's technology office confirmed in September 1999 that the project had been "postponed."[1] Its future now lies

in the hands of the new government that came to power in October 1999, which is likely to be far more skeptical than its predecessor. In his first week in office, President Wahid heaped scorn on another of ex-President Habibie's hi-tech megaprojects, the development of an Indonesian aircraft industry,[2] and is likely to take a dim view of the Mamberamo scheme as well. But as the Kalimantan PLG megaproject has shown, uncontrolled land clearing for a project, even one that eventually fails and is abandoned, can fundamentally change the ecosystem and make extensive fires and other forms of degradation a virtual certainty. It can only be hoped that the lessons of the PLG project will be taken seriously and that this ill-advised scheme to pillage one of Indonesia's last major blocks of pristine forest will be aborted before it does too much damage.

Source:
Adapted from Carr, 1998.

Notes:
1. Agus Sugiyono, personal communication, September 21, 1999.
2. "Indonesia's Wahid Scorns Habibie Hi-Tech Dreams." *Reuters*, October 26, 1999.

The PLG megaproject is only the most egregious of many similar projects that have degraded Indonesia's forests over the past three decades and have provided the tinder and the spark for the worsening cycle of megafires during that period. Indeed, an even more implausible megaproject has been planned for Irian Jaya's 7.7 million ha Mamberamo River basin. (*See Box 10.*) The suffering and losses of the Mengkatip River Dayak people are unique only in that the situation has attracted a larger than usual share of attention from journalists and environmental activists.[195]

> *Look, if we don't change our ways, we won't survive as a nation, all right? I hope by this time it's clear to everybody.*
>
> Sarwono Kusumaatmadja
> Minister of State for Environment
> October 6, 1997

The project is emblematic of Suharto-era policies and their impacts on forest ecosystems and forest-dependent communities, and it neatly sums how flawed development policies in the hands of an authoritarian and unaccountable government, riddled with corruption, set the stage for fire disasters like that of 1997–98. The steps that the government of President Wahid takes, or fails to take, to make amends for this disaster in the heart

of Borneo, and whether it cancels similar disasters in the making such as the Mamberamo megaproject and the proposed "Kakab" successor to PLG, will serve as a useful test of the extent to which the reforms discussed in the next chapter are being carried out over the coming years.

Reversing 30 years of ill-considered forestry and land-use policies and repairing the damage they have done to Indonesia's forests is a daunting task. Much of the damage can never be repaired within a time scale meaningful to humanity. The great lowland forests in much of Kalimantan and Sumatra, for example, are gone for good, and countless species have been driven to extinction in the process. The peat-swamp forest areas devastated by the million-hectare rice debacle will remain, for all practical purposes, a vast, grim monument to "a crackpot, Stalinist-style plan to reorder nature" that was "certain to fail because of its unmanageable scale and the unforgiving, little-understood peat terrain."[196]

An important place to start is with immediate measures to reduce fire hazards and fire risks and strengthen firefighting capacities before the arrival of the next extended drought and burning season, which is likely to be associated with the next El Niño. Ninety-three percent of all droughts in Indonesia between 1830 and 1953 occurred during an El Niño event, as have most of the droughts since then, including those associated with the extensive fires of 1982–83, 1994, and 1997–98.[197] Recent experience seems to indicate that El Niño events are occurring more frequently. The exceptionally long El Niño event that began in 1991 peaked only in 1994 and did not end until early 1995, for reasons not yet well understood.[198] This was followed, just two years later, in 1997, by one of the strongest El Niños ever recorded. If this El Niño frequency continues, Indonesia could begin this new millennium in flames once again unless preparatory measures are taken.

Taking immediate action to address the proximate causes of Indonesia's periodic infernos is in no way inconsistent with addressing the much broader agenda of reforms needed to deal with the root causes of the fires and other key forest degradation processes identified in Chapter VI. The broader policy reform agenda needs to be carried out in phases, as the World Bank suggests.[199] Some steps can be and need to be taken immediately to secure the remaining forests from pressures that are intensifying as a result of the economic crisis and to respond to the strong and growing political demand for reforms giving indigenous and other forest-dependent communities greater access to the benefits that forests provide. Other steps must be taken immediately because of government pledges to the IMF as a condition of the economic bailout package. (*See Box 11.*)

Other reforms will take a good deal more time, in part because the issues are so complex and the changes being discussed are so sweeping. Moreover, these major policy and institutional shifts can succeed only through an open and transparent process of public dialogue among competing interest groups—a political phenomenon that Indonesia has not experienced in over three decades, if ever. And finally, Indonesia must get its political house in order so that forest policymakers who possess the mandate, political credibility, and long-term vision can carry out a reform program. The government that took power under President Wahid in October 1999 has strong reformist credentials, but its capacity to actually implement forest and land-use policy reform is untested and unknown as of this writing.

Is Real Reform Possible?

A 1998 World Bank memorandum on forest sector reform acknowledged that for the first time since 1966 a consensus in favor of sweeping reform has indeed formed among elements of the government, the private sector, donor institutions, and many NGOs. But neither the old systems of power and privilege nor the actors who benefited from them have left the scene.

"There remains a strong element in the forest industry, and in the official forestry agencies, that will resist reform, or at best will give it token acceptance while attempting to preserve the privileges of the past. All that can be said, at this point, is that the political predominance of an industry based on vested interest and institutionalized market distortion can now be seriously dealt with, in a manner that was not previously possible." [200]

Is substantial reform really possible? One can only answer that the prospects for reforming the forest sector are better than they have been for three decades but that reform is by no means assured. Suharto has been ousted from office, but much of his regime and the people who have run and profited from it—from Jakarta to the most remote villages—are still in place. Old-guard forestry bureaucrats and the greater part of the industry oppose the *reformasi* movement to the extent that it aims to weaken their power and diminish their profits, and "both these groups still retain great control—both de jure and de facto—over decisions determining the disposition and conversion of forest areas."[201]

At the same time, the economic crisis has lent renewed urgency and legitimacy to policymakers' calls for intensified short-term exploitation of the country's natural resources, to provide both food security and export income. This dynamic creates further barriers to forest policy reforms that are accused of slowing recovery from the crisis and of hampering economic growth.

In the countryside, the significant proportion of the populace that cannot benefit from favorable commodity terms brought about by currency devaluation is not waiting for policy pronouncements on the topic. Facing massive layoffs in the urban manufacturing and service sectors, and drastic rises in the prices of basic commodities, millions have turned to the forest as a ready source of income. For example, illegal capture and export of wildlife, driven both by economic need and by high export prices, has become an epidemic unlike anything seen in the country in the past. "Now, it's back to every species for itself," in former environment minister Emil Salim's words.[202]

The economic crisis has also crippled the government's already weak capacity to supervise and monitor logging and plantation operations and enforce forestry laws and regulations. Department budgets have been slashed across the board, while the cost of travel has increased. Simultaneously, the level of respect for (or fear of) the law—or at least of those who enforce it—that existed in the Suharto era has largely evaporated, owing, in great part, to the low esteem into which the Suharto regime brought "the rule of law."[203] The current situation presents parallels to what happened in the then-rich teak forests of Java when the Japanese conquered the island in 1942:

11. FOREST POLICY REFORM CONDITIONALITIES IN THE IMF ECONOMIC BAILOUT PACKAGE

As Indonesia's economy crashed in 1997, the government opened negotiations with the IMF for a financial assistance package. Agreement was initially reached in November 1997 on a wide-ranging set of basic economic and fiscal reforms that the government agreed to undertake in return for assistance totaling about $40 billion. The assistance was to be provided by a consortium of donors, including the IMF, the World Bank, the Asian Development Bank (ADB), and various individual governments, notably Japan.

The policy conditionalities and their deadlines have been repeatedly revised since then. The changes reflect the roller-coaster nature of Indonesian political and economic developments and, according to some observers, the flip-flops of the IMF and other donors, who steadfastly stuck to economic analyses even though the mixture of the economic crisis and the country's volatile politics was clearly the key dynamic that needed to be dealt with. (*See Box 2.*)

For one of the first times in its history of prescribing structural adjustment measures in return for emergency infusions of capital, the IMF (and its partner institutions) in January 1998 prescribed among the conditions for the bailout a number of forest policy reforms including:

■ An increase in the forest land tax.
■ Direction of inflows to the Reforestation Fund (collected as a production levy on logging operations) to the official government budget, rather than retention under the unmonitored control of the minister for forestry and other political leaders. Fund monies had been misallocated for numerous nonreforestation projects, including the state aircraft corporation, which former President Habibie previously headed.
■ Abolition of existing forestry levies and their replacement by a resource rental tax.
■ Removal of the restrictive forest products marketing arrangements embodied in APKINDO, the exporters' cartel run by one of Suharto's cronies, Mohamed "Bob" Hasan.
■ Reform of logging concession regulations to allow for periodic review of stumpage charges, lengthening of concession terms beyond the current 20-year limit, and authorization to trade concession rights. The latter two provisions were intended to give concessionaires a commercial incentive to practice better forest operations and management.
■ Competitive auctioning of concession rights.

In April 1998, the World Bank followed up with further policy reform requirements. Specific measures proposed in the loan for the forest sector (supplementing or elaborating on those in the IMF program) were:

■ linkage of forest royalties to world prices;
■ reduction of export taxes on forest products to 30 percent ad valorem immediately and to 20 percent by the end of 1998;
■ introduction of an independent system for monitoring forest resources, including participation of local communities, by the end of 1998;
■ a moratorium on issuing new logging licenses until these new measures are in place;
■ introduction of performance bonds on forest operations; and
■ development of sustainable forestry land management targets.

Some progress has been made on implementing these conditions, but for most, the very tight deadlines have not been met. In general, Indonesia is not to blame; the unrealistic deadlines arose from the political pressures on the IMF by the governments that fund it to specify benchmarks on which to base disbursement of successive tranches of funding. The IMF was understandably concerned to release funds as soon as practically and politically possible to forestall further meltdown of the Indonesian economy and the feared regional and global "contagion" effects. But donor governments, sensitive to political resistance to using taxpayer money to bail out faraway foreign countries, demanded tangible indicators of reform by the recipient government. For many of the conditions, such as setting up

a banking restructuring agency, a short time frame was workable. Forest policies, however, are complex, and many of the actions mandated by the IMF, such as auctioning concessions, have no precedents in Indonesia. Thus, the deadlines for many of the forest policy-related conditions have been allowed to slip, although pressure to complete the reforms continues.

It is worth noting that another set of IMF-mandated reforms lifts export and external investment restrictions on oil palm, a move that is sure to increase forest conversion. In a recent policy memo the World Bank distanced itself from these measures, referring to them as "IMF-originated policies," but went on to say that those policies are not intended, and should in no way be interpreted, to mean that viable natural forest areas should be converted to oil palm. This is perhaps a well-intended argument, but in fact those IMF-mandated policies are indeed intensifying pressures for conversion of natural forests, and there are virtually no restraints on such conversion in the current policy environment.[1]

Source:
World Bank, 1998b.
Note:
1. Potter and Lee, 1998a.

"The Dutch colonial government in Java, and the mystique that had permitted it to rule for nearly 150 years, fell within ten days of the Japanese invasion. Forest villagers believed that the end had come to the restrictions keeping them out of the forest [and they] responded vehemently to the sudden change in the forest custodians. They ransacked remaining logyards, administrators' housing, and the forest itself." [204]

In an eerie echo of those violent times, mobs in East Java began looting the government-owned teak forests around several villages in August 1998. In the police operation mounted in early September in response to the mass timber thefts, one villager was killed, several were seriously injured, and hundreds were forced to flee their homes.[205] Similar events are occurring increasingly frequently across the archipelago, fueled by economic

desperation, anger at the government, and, in many cases, opportunistic exploitation of the situation by well-organized gangs of full-time looters, with all parties using *reformasi* as the justification for their actions.[206] The legitimacy of Suharto-era local officials has been called into question across the country, and thousands of village heads have been forced to step down.[207]

Rebuilding the legitimacy and capacity of the government bureaucracy at the local level will clearly be a long and troubled process, but such renewal will be essential for effective implementation of forest policy reforms at the local level.

Despite these considerable problems, there are real signs of hope that effective forest policy reform may indeed be possible in the post-Suharto era. Already, a number of broad-based groups have been formed to debate forest policy issues and devise agendas for change.

In June 1998, the minister of forestry and estate crops issued a decree establishing the Committee for the Reform of Forest and Estate Crop Development, a group composed of officials, academic forestry experts, forest industry representatives, and several environmental NGOs. The committee's mandate is sweeping, covering the formulation of a broad forest policy and an institutional reform agenda and monitoring of progress in implementing reforms.[208] It is unclear to what extent the reform-minded views of its members represent the general views of forestry officials: at least one member and a number of observers have expressed frustration that recent official decisions and policy moves bear little resemblance to the committee's recommendations and instead more or less perpetuate existing policies,[209] and one NGO member has resigned in protest. But the very existence of this body is a considerable departure from past practice in the ministry.

Another group, the Communication Forum for Community Forestry (FKKM), an independent group of academics, NGOs, forestry officials, and some donor agencies, met for the first time in June 1998 and produced a statement articulating the outlines of a reformist vision. The statement, which was delivered to the minister of forestry and the parliament, argues that for reform to succeed, .

"First of all, parties and stakeholders involved in the implementation of national forest management should realize that the condition of the nation's forests is now very poor, as indicated by increasing areas of cleared land, land disputes, poor spatial planning, low productivity, limited access for local communities, and lack of government recognition of local communities' right to utilize the forest." [210]

In his report on the June 1998 meeting, the Forum's chair stated that:

"During the last three decades, management of Indonesia's forests has neither contributed to the people's welfare, especially local and indigenous peoples, nor has it guaranteed the conservation of forest resources. The government has consciously deviated from its constitutional mandate to manage forest resources sustainably and allocate as much as possible for the prosperity of the Indonesian people. The current forestry crisis is not merely the result of mismanagement, but rather is none other than the result of the government's adherence to the wrong paradigm of forest management."

Rebuilding the legitimacy and capacity of the government bureaucracy at the local level will clearly be a long and troubled process, but such renewal will be essential for effective implementation of forest policy reforms at the local level.

His statement went on to note the "criminal" misappropriation of money from the Reforestation Fund to destroy vast areas of forest for the million-hectare rice project in Central Kalimantan and argued that:

"This carelessness demonstrates to us that the government has promoted stupid development processes which cause disasters both in terms of forest resources and suffering for the generations to come. This catastrophe is the result of both expert consultants and the government apparatus adhering to inherently flawed knowledge which has also resulted in the economic crisis, political crisis, food crisis, forest fires crisis, and moral crisis."

It is worth noting that as recently as early 1998 it would have been unthinkable for any but the most daring environmental activitists to present publicly this kind of rhetoric—which is now coming from mainstream academics in the leading forestry schools and from some government officials themselves.

The 1997–98 fire catastrophe has lent additional momentum to reform. The fires marked the first time that the government officially acknowledged the link between the fires, with their disastrous effects, and the actions of private firms in the forestry and plantation sector—a link that was widely reported and condemned within the country. In addition, the sharp international reaction to the haze that spread across the region embarrassed the government. Most significant perhaps, was that ASEAN took up the haze problem as a legitimate regional issue, without objection from Indonesia. As fires again began to burn in mid-1999, ASEAN member Brunei, concerned about the threat to the Southeast Asian Games it was about to host in August, threatened to sue Indonesia if it did not control fires on Sumatra and Borneo. [211]

Adding a realpolitik element to the pressure for forest policy reform is the package of forest sector reform measures agreed to by Indonesia as part of the massive IMF bailout package. (*See Box 11.*) The IMF, the World Bank, and the ADB—the three main partners financing the recovery initiative—need to be sensitive about seeming to bully Indonesia and raising nationalistic hackles. (Tommy Suharto, the former president's son, for example, called the agreement with the IMF "neo-colonialism."[212]) But there can be no doubt that Indonesia must depend on these institutions to keep the country from spiraling into total financial chaos and that they therefore have immense power to influence various policy choices, comparable to their power in the early days of Suharto's rule in the late 1960s.

A key element for success will be to engage and enlist the private sector in the process of reform. The value of Indonesia's forest-based exports is expected to top $8 billion for 1999,[213] and the timber, pulp, and paper industries, as well as the fast-growing oil palm sector, will be an important part of the Indonesian economy for the foreseeable future.

> *The international community is giving Indonesia a hand in its recovery from the Asian financial crisis. The forest fires, and the underlying economic and political policies, should be on the international agenda. The International Monetary Fund and the World Bank must see this as an issue of unjustified subsidies, unsustainable development and poor governance that is clearly withinin their mandate.*
>
> Simon S.C. Tay, Chairman
> Singapore Institute of International Affairs
> *International Herald Tribune,*
> August 31, 1999

Most of the hundreds of logging and plantation firms hold legal contracts with the government to operate their concessions. Although it would be possible for the current government to abrogate these contracts, it is not politically or economically feasible, at least in the short term, and would most likely result in a massive number of court cases that could drag on for years. In any case, it is a much wiser course to provide firms that have existing concessions and are obeying the terms of their contracts with an opportunity to carry out a transition to more environmentally and socially sensitive ways of operating, in line with reform policies as they evolve.

Such an approach does not mean that companies that are violating their contracts and degrading the forest should not have their licenses revoked—something entirely within the law and the terms of their agreements. But the brighter prospect lies in assisting progressive companies to change their practices toward a more sustainable and equitable model of forestry. The recently formed Indonesian Ecolabeling Institute (LEI), the country's official timber certification body, will be a key institution in drawing the private sector into the reform agenda. (*See Box 16.*) Already, a number of logging companies are seeking LEI certification and are cleaning up their practices to that end.

As of mid-1999, a considerable number of new forest laws and regulations had been passed or were under discussion, including a new Basic Forestry Law, passed by Parliament in September 1999, and a new government regulation on logging and timber plantation concessions, passed in early 1999, that included a number of the provisions required by the IMF bailout conditions. An analysis by a member of the Forest and Estate Crops Development Reform Committee, however, asserts that these various regulations were hurriedly passed to meet the IMF conditions and that forestry bureaucrats freely admit they will have "no implementation consequences."[214]

The new Forestry Law has provoked widespread condemnation and opposition from, among others, a coalition of 125 NGOs and two former ministers. Djamaluddin Suryohadikusumo, a former forestry minister, argued that "no part of this draft recognizes or protects the rights of local tribes living in the forests" and that the bill "will not change the mind-set adopted by our timber companies of exploiting the forests to the maximum," noting that logging concessions had destroyed nearly 17 million ha of forests over the past three decades. Former environment minister Emil Salim also spoke out against the bill, arguing that it would "result in the rampant felling of trees in protected forests and cause a boom in the illegal timber trade."[215] In another interview, Salim said, "The whole law is very much government controlled, very much top down. Where is the role of the people? Where is the role of civil society? It's not there." A World Bank official in Jakarta said that the law did not fulfill the reforms required by the Bank as a condition of an economic bailout loan approved in May and pointed out that recommendations by the National Forest and Estate Crops Development Reform Committee had been ignored. The official added, "We have been urging [the government] to set up some kind of consultative body within the Forestry Department but they obviously haven't."[216]

With a new parliament and president in power since October 1999, however, the fate of the new forestry law is in doubt. A wholesale revision by the new parliament, in a more reformist direction, is possible, especially if Indonesia changes its constitution to become a federal state, something that President Wahid has spoken out in favor of. [217] A more likely scenario in the short term, however, is that reformist pressures will lead to issuance of implementing regulations that encompass a significant proportion of the forest policy reform agenda promoted in 1998-99 by the various forest policy reform committees and groups noted above. The new law is certainly vague enough on many points to allow for a significant degree of creativity in its legal elaboration and its implementation in the field.

All in all, the climate for forest policy reform is better than it has been in more than three decades. But, as the World Bank assessed the situation in its presentation to the July 1999 meeting of the donors' Consultative Group on Indonesia (CGI):

"The political changes of 1998 and 1999 have resulted in very important changes for forest resources. Government has put unprecedented energy into forestry policy reform over the past year, but there has been inadequate consultation and acute uncertainty persists. This uncertainty amplifies the risk for forest resources because it induces further exploitative activity." [218]

The next few years thus provide an unprecedented window of opportunity during which the new government, NGOs, reform-minded elements of the private sector, and the international community must act.

STABILIZE, LEGALLY PROTECT, AND DEFEND THE REMAINING FOREST ESTATE.

■ CARRY OUT AN ACCURATE INVENTORY OF VEGETATIVE COVER AND LAND USES LYING WITHIN THE LEGALLY DEFINED FOREST ESTATE.

One of Indonesia's immediate priorities is to complete an accurate inventory, using both spatial and statistical methods to present the data, of the vegetative cover lying within the 143 million ha of land officially designated as state forestlands. The technical difficulties are not great, and some of the work has already been carried out. A national forest inventory was completed in 1995, but the government has never officially released the full results. There are, however, significant gaps and needs for updates. Existing data are scattered among various projects and offices; and significant data collected by the timber industry have been withheld from the public and even from the Ministry of Forestry, in some cases. The World Bank-assisted national forest cover mapping effort carried out in 1998-99, discussed above, should provide a useful baseline. The government should make an immediate and nonnegotiable demand that the private sector publicly release, at its own expense, information on the timber industry. In turn, the government should make the inventory public, in forms useful to the academic community, the media, NGOs, and the citizenry. Donor agencies and NGOs should lend their support to the publication of the inventory and its dissemination to as broad an audience as possible.

■ GRANT CLEAR LEGAL PROTECTION AS PERMANENT FOREST ESTATE TO ALL REMAINING FORESTED AREAS.

On the basis of the inventory, all remaining forest areas should be given unambiguous legal protection as permanent forest estate not available for conversion to other uses (such as timber and oil palm plantations) except in unusual circumstances and through a transparent and accountable decisionmaking process. Conversely, an accurate accounting of areas that are available for conversion to other uses (truly degraded forestlands and lands already stripped of forest but still classified as forest) needs to be carried out as part of this process.

■ STABILIZE KEY PROTECTED AREAS.

Stabilizing and defending the boundaries of those protected areas that are most important in preserving representative samples of Indonesia's globally important biodiversity must be given a high priority. By March 1998, Indonesia had (on paper) established 36 national parks (14.5 million ha), 177 smaller Strict Nature Reserves (2.4 million ha), 48 Wildlife Sanctuaries (3.5 million ha), and a variety of smaller recreational and hunting parks totaling about 1.3 million ha, a total of 21.7 million ha. An additional 34.6 million ha were designated as Protection Forests due to their watershed values, steep slopes, or fragile soils. [219] Thus, some 56 million ha of the country's forest lands—more than a quarter of its land area—are in theory off-limits for any activities that degrade or remove their forest cover.

Unfortunately, only a very small percentage of this vast area is effectively protected. Most parks and protection forests are subject to pervasive encroachment for small-scale agriculture, conversion (legal or not) to plantation crops, illegal logging, wildlife poaching, and mining. Even large, well-known parks such as Kerinci Seblat and Leuser in Sumatra—which together represent the last relatively pristine large forest areas on the island and have been supported with millions of dollars in international aid—are being rapidly degraded.

Halting the degradation of all these areas should be the ultimate goal of Indonesian forest policy, but in the short term, this is realistically impossible. Rather, as the World Bank has recommended, the government should give priority to a limited number (the World Bank suggests 10) of protected areas that contain the country's largest, relatively undisturbed expanses of forest and should initiate an intensive campaign, in collaboration with international and national conservation NGOs, to raise international funds for stabilizing the boundaries of these areas and developing effective protection regimes. Conflicts between protected areas and local communities are common throughout the country, and efforts to reconcile community and conservation efforts have met with mixed success. (*See Box 12.*)

12

INTEGRATED CONSERVATION AND DEVELOPMENT PROJECTS IN INDONESIA

It has long been recognized that traditional western models of protected areas management—the "Yellowstone" model in which all human economic activity is forbidden and penalized—are counterproductive in countries such as Indonesia where numerous forest-dependent communities commonly live near park boundaries or in enclaves within protected areas and have often occupied those areas longer than the protected area has existed. Accordingly, most recent conservation projects in Indonesia have followed the integrated conservation and development project (ICDP) model.[1]

Unfortunately, the record of ICDP approaches to slowing degradation of protected forests in Indonesia has not been good. A 1997 report commissioned by the World Bank concluded that "very few of the ICDPs can realistically claim that biodiversity conservation has been or is likely to be significantly enhanced as a result of current or planned ICDP activities. . . . [M]any or the most immediate problems faced by ICDPs reflect flaws in the basic assumptions and planning, which are not well-matched to the real threats and capacity constraints that conservation projects face in the field." [2]

Part of the problem with the ICDP model has been an uncritical acceptance of the notion that local community use of forest resources and protection of forest biodiversity can always coexist. This has led to an overemphasis on development of local economic activity in "buffer zones" around protected areas, an approach grounded in the dubious assumption that creating intensified local economic activity on the borders of parks will somehow keep people out of it, rather than draw more people into the area.

At the root of the problem in Indonesia is the fact that "parks and people" have been pitted against each other in a struggle over the small remnants of forestland that have not been taken over by commercial private sector interests allied with the government—principally logging, plantation, and mining concessions. If access to these lands currently occupied by private firms were to be shared more equitably with local communities, as would be the case under the "community concession" model recommended in this paper, pressure on protected areas could be appreciably reduced. This is the most promising strategy for resolving conflicts between local communities and priority protected forest areas, although buffer zone approaches still have a significant, if subsidiary role.

In addition, the government and the donors supporting its forest conservation efforts need to put renewed emphasis on some of the more traditional fundamentals of protected area management. The World Bank study cited above concluded that "the largest obstacle confronting ICDPs on the ground has been the lack of PHPA [Park Service] capacity," and stated that stengthening the PHPA is clearly a key to more effective forest protection. The study went on to note, however, that recent "foreign technical assistance and institutional support have tended to substitute for capacity development rather than to produce it." Building real capacity to police park boundaries and punish poachers and illegal loggers—tasks that foreign consultants or NGOs cannot carry out—is a high priority.

Notes:
1. For a review of the theory and practice of ICDPs, see Wells and others, 1992; for an analysis of ICDPs in Asia, see Barber, 1995.
2. World Bank, 1997.

RECOGNIZE AND LEGALLY PROTECT FOREST OWNERSHIP AND UTILIZATION BY INDIGENOUS AND FOREST-DEPENDENT COMMUNITIES AND ASSIST THEM IN MANAGING THE FOREST SUSTAINABLY AND PRODUCTIVELY.

Once the true forest is legally secured, a process of reordering its uses—and users—can begin in earnest. And once there is an accurate accounting of unclaimed degraded forestland available for other uses, decisions can be made on the most efficient and equitable distribution of those areas among various stakeholders. But before any zoning or allocation takes place on these lands, the long-standing wrongs committed by the Suharto government against the rights and livelihoods of indigenous and other forest-dependent communities must be corrected.

Redressing the continuous erosion of local and indigenous community access to, and use of, Indonesia's forests has long been a key objective of the Indonesian and international NGO community. The Suharto government steadfastly refused to acknowledge the customary rights of Indonesia's numerous indigenous forest-dwelling peoples (and, indeed, denied that Indonesia had distinct, minority indigenous peoples) or to recognize the plight of the millions of other forest-dependent local people impoverished by its logging, plantation, transmigration, and mining policies.[220]

In the aftermath of the collapse of the Suharto regime, a broad spectrum of reformers is arguing that a reordering of the relations between the government, local and indigenous communities, the private sector, and the forest is a central element of a more just and sustainable forest policy. The World Bank, for example, maintains that at least 30 million people are highly dependent on forests for important aspects of their daily livelihood, that the economic crisis is likely to increase their numbers, and that any workable forest sector reform agenda "must give primacy to radically increased participation of forest-dwelling and adjacent communities in the management, utilization, and actual ownership of forests and forested lands." [221]

For its part, the government fears that if alternative sources of livelihood cannot be developed for a burgeoning and increasingly desperate rural population, further political chaos and civil violence may be sparked by rising unemployment in the manufacturing and services sectors, combined with rising prices for basic goods. Granting local communities greater access to forest lands and resources may thus be a tool for the government's political survival, as well as a way to visibly respond to the growing clamor for *reformasi*.

> *If the state will not recognize us, we will not recognize the state.*
>
> Preamble, Decisions of the First Indonesian Indigenous Peoples' Congress
> Jakarta, March 21, 1999

Key actions that need to be taken on the customary ownership issue include the following:

- **LEGALLY RECOGNIZE OWNERSHIP OF FORESTS LYING WITHIN THE CUSTOMARY TERRITORIES OF INDIGENOUS AND TRADITIONAL (*ADAT*) COMMUNITIES.**

Indonesia's forestry laws and regulations should explicitly recognize the principle that traditional (*adat*) communities own the forest areas within their customary territories and have the right to utilize them sustainably, provided that the areas are maintained as permanent forest estate. To this end, a process should be established whereby *adat* forests are mapped and a written agreement is concluded between the Ministry of Forestry and the traditional or indigenous community, represented by the leaders of their customary institutions of governance. (*See Box 13.*) The agreement should carefully specify the obligations of the community with respect to maintenance of the forest (*see Box 14*), explicitly affirm the government's recognition of the community's rights, and pledge the government's support in defending the forest against encroachment and exploitation by actors from outside the community.

13

COMMUNITY MAPPING STRATEGIES AND TECHNIQUES

Mapping forest areas is an intrinsically political act.[1] Official Indonesian forest maps establish the territorial claims of the state over 74 percent of the country's land area and demarcate the subordinate claims of a variety of concession holders who have close ties to the state apparatus and are engaged in natural resource extraction.[2] These maps exclude the settlements, resource uses, and traditional claims of local communities to forest lands and resources. The unmapped uses and claims, overlaid with the official maps, pinpoint the myriad conflicts over forest resources that have plagued Indonesia for the past three decades. These conflicts must be resolved if forest management is to be made more sustainable and more equitable. Utilizing community mapping techniques and integrating them into land-use planning, allocation, and management of forestlands is an important avenue for progress in this regard.

Community-level sketch-mapping has been widely utilized for some time in many countries as a tool for rapid rural appraisal, community forestry efforts, and advocacy on behalf of traditional land claims against external threats. The advent of inexpensive and simple global positioning system (GPS) technology has made it possible for such local mapping exercises to be georeferenced with national mapping methodologies. Many successful examples exist around the globe[3] and in Indonesia.[4] In their efforts to challenge state forest land-use allocations that ignore their own claims and interests, local communities now have the tools to speak the language of dominant mapping systems and thereby challenge them.

Traditional communities in Indonesia can, first of all, use this technology in support of efforts to gain basic recognition of their rights over particular forestlands. More than a map is required for this, of course—there needs to be some showing of long-term occupancy and use, for example—but without a georeferenced map, traditional land claims remain indeterminate and difficult to press. Under the Suharto regime, even a well-made, georeferenced map supported by well-documented claims of long-term traditional forest occupancy and use was unpersuasive in the face of unrelenting government hostility to recognizing such claims under any circumstances. But times have changed, and some level of formal state recognition of traditional forestland claims appears inevitable in the near future.

Once such claims are accepted, there remains the problem of demarcating boundaries on the ground. Older surveying techniques were so slow and expensive that they were effectively beyond the reach of almost all local communities. Indeed, the Indonesian Ministry of Forestry, riding one of the world's largest and most profitable timber booms, has been unable to demarcate most of its own claims on the ground.[5] GPS technology promises to make demarcation far less expensive and time consuming, and it is a simple technology that can be easily taught. The terrain in question is a factor, of course: "Rivers make for fast work, while forested mountains slow the process down."[6]

For community mapping to become more than a sporadic pilot project phenomenon, the government's regulations on forest boundary demarcation must change. A new draft regulation on establishing and demarcating forest boundaries was under

discussion in early 1999,[7] but it is mostly concerned with the decentralization of functions from central to provincial and district government units. It makes no mention of negotiating boundaries with local communities. Furthermore, it maintains the long-standing requirement that boundaries be marked with concrete posts of a certain size and dimension before a forest boundary is officially demarcated.[8]

Even if the regulations can be reformed so as to accept and integrate GPS-based community mapping and demarcation, a great deal of work is needed to build mapping capacity at the community level. As Peluso (1995) notes:

"While counter-mapping has some potential to transform the role of mapping from a "science of princes," it is unlikely to become "a science of the masses" simply because of the level of investment required by the kind of mapping with the potential to challenge the authority of other maps. . . What ultimately may be more important for the "masses" is not the technology itself, but the content of the maps produced and the way the knowledge and information on the maps is distributed."

Community mapping is not a panacea, but it is an increasingly important tool for establishing secure local claims over forest resources. And securing these claims is an important prerequisite for reducing conflict over forest resources and providing incentives for their sustainable management. The challenge in Indonesia is to both build community mapping capacity and reform government policies so that the results of community mapping become a part of forest policy rather than a challenge to it.

Notes:

1. "Forest maps pinpoint the location of valuable and accessible timber and mineral resources . . [and] have been an important tool for state authorities trying to exclude or include people within the same spaces as forest resources." (Peluso, 1995).
2. The first comprehensive forest-mapping exercise in Indonesia was the development, in 1981–85, of provincial consensus forest land-use plans (Tata Guna Hutan Kespakatan, or TGHK) that divided the forest estate into various categories such as production, protection, and so on. Developed from old data, mostly without verification on the ground, these maps not only excluded all community claims and uses but sometimes placed whole towns within protected forest zones. A late 1980s effort, the Regional Physical Planning Program for Transmigration (RePPPProt), was developed from satellite and aerial imagery to determine suitable locations for new transmigration sites and associated plantations. Although these maps dramatically improved the representation of vegetative cover, they still did not include data on local forestland uses and claims. (Ibid.)
3. See, for example, the 50 cases discussed in Poole, 1995.
4. Momberg, Atok, and Sirait, 1996.
5. In 1996, the Ministry of Forestry estimated that of the 352,000 km of state forest boundaries (both outer boundaries and boundaries between functional categories) that needed demarcating, only 113,594 km (32 percent) had actually been demarcated by 1994 (Ministry of Forestry, 1996). The remainder, more than 238,000 km, is more than five times the circumference of the Earth.
6. Poole, 1995.
7. Ministry of Forestry and Estate Crops, 1999.
8. Poole (1995) cites a boundary demarcation effort in the territory of Brazil's Kayapo tribe where the most expensive item in the $600,000 budget was the use of helicopters to transport cement for boundary markers required by Brazilian regulations.

TRADITIONAL RESOURCE RIGHTS AND FOREST CONSERVATION

Proposals—whether by governments or by environmentalists—to recognize or compensate traditional forest claims are usually qualified by the assertion that such recognition or compensation should be part of a *quid pro quo* arrangement in which the community agrees to certain conditions and guidelines for "sustainable resource management." For their part, advocates of indigenous rights over forests have often asserted that recognition of such rights will invariably lead to forest conservation, since indigenous people's traditions predispose them to sustainable management. (This argument has been very successful in recruiting environmental activists for the indigenous rights cause.) As indigenous rights over forests are gradually recognized in various parts of the globe, however, the argument is increasingly heard that "ownership is ownership:" if a community's traditional claims to a forest area are indeed valid, then the community has the right to do as it pleases with the area, regardless of the impacts on biodiversity and other factors valued by outsiders.

It is understandable that traditional communities in Indonesia would be skeptical of such restrictive arrangements: for decades, they have watched while the state parceled out their territories and resources to outsiders who plundered timber and other resources without regard for "sustainability" and without interference by the state. Now, suddenly, just as the state decides to recognize long-standing local claims, it puts forward a whole series of restrictions on those claims. There is no short-term solution for this problem: it will take years of good-faith actions by the state to help traditional forest communities overcome the legacy of mistrust.

But the assertion that "ownership is ownership" is a red herring. Whether one looks to western systems of property law or to Indonesia's own rich legacy of traditional *adat* property law, there are numerous shades and varieties of "ownership" over land and resources. Property rights may be bounded in time, restricted to certain uses, and limited in many other ways. And everywhere, the exercise of property rights is limited by

considerations of public interest. One may hold full title to a house and land, for example, but not have the right to establish a toxic waste facility in the front yard. Similarly, a traditional community might be granted a strong property right over its local forests but not the right to clear-cut watershed slopes, set fires during droughts, or exterminate legally protected species of fauna and flora.

For some indigenous communities with distinct cultures and territories apart from and predating the dominant culture and state system—and many of these exist in Indonesia—the issue is not the legal issue of "property rights" but the political issue of "sovereignty." The rights that communities claim in such cases are more like those of a "state within a state" than a normal property right. That is, they seek the autonomy not only to "own" their territory but also to be the legitimate political and lawmaking authority within that territory and in external relations: "The essence of *hak ulayat* [traditional sovereign rights over territory] lies in 'autonomy' and/or 'sovereignty.'"[1]

These are demands that the Indonesian government needs to consider seriously if it wishes to restore trust and civility to the currently poor relationships between the state and forest-based indigenous communities. Very few communities, however, want to completely cut themselves off from the modern economy and the dominant political system and culture; rather, they seek to recapture control over their traditional resources and territories and to ensure that their engagement with the dominant culture and economy is within their control rather than forced on them. Most forest-dependent communities in Indonesia are not discrete, isolated cultures seeking a "state within a state." They seek, instead, recognition of their claims over resources that are integral to their economies and cultures, respect for their cultural traditions, and protection from outsiders who threaten these things. If the Indonesian government can meet those expectations, negotiating for sustainable management of forest resources should not be so great a challenge.

Note: 1. Zakaria, 1999.

■ **ESTABLISH A NEW "COMMUNITY FOREST CONCESSION" RIGHT THAT MAY BE GRANTED ON STATE FORESTLANDS.**

Not all (or even most) forest-dependent communities in Indonesia possess the long-standing connection to a particular forest area that indigenous and traditional peoples do. For these groups, a community forestry concession right should be available to legitimize their existing activities on state forestlands (where those activities are sustainable) and to provide them with long-term incentives for serving as good stewards

of the forest and for carrying out long-term activities such as tree planting and agroforestry on degraded lands. These contracts would be legally similar to the concessions currently given out for commercial logging and plantation operations, in the sense that they would be for a fixed, renewable period of time and would clearly specify the rights and responsibilities of the concessionaire. Recipients might be an organized group of "forest farmers," a family, or a whole community. As with timber concessions, the contracts should specify permissible uses of forest resources and establish criteria and

monitoring systems to ensure that the terms of the contract are met. As discussed below, they should not be restricted to timber exploitation.

While recognition of, and support for, local and indigenous forest access and use are increasingly seen as key elements of an effective reform strategy, it is dangerous to romanticize the prospects for ecologically sustainable local management, even by relatively isolated traditional communities, in a context of pervasive global markets, ubiquitous demand for modern consumer goods, and economic crisis. Vayda (1998) warns against "regarding local people's

control, by itself, as a virtual panacea for environmental problems" and notes that "gaining control over long-term management of a resource may lead local people, especially if they have had the past experience of booms and busts in particular forest products, not to conservation-oriented management but rather to their own intensive exploitation of the resource as long as it fetches a high price and remains fairly readily available." Sanderson and Bird (1998) have similarly warned against the "magic of tenure" notion, whereby giving particular people greater control over particular resources is assumed to "guarantee better environmental outcomes." And the World Bank, while supporting a "radical" shift in the direction of community access and control, cautions that

"[T]he complexities of entitling communities to forests, no matter how justified and urgent, cannot be rushed: even in countries where community title to large forests is an undisputed fact and has been in existence for years, or even centuries, the complexities, disputes and failures to benefit some people within the community groupings have been serious and potentially destructive of the whole idea." [222]

With those caveats in mind, it is important that both the new legal framework for recognition of community forest rights and the process by which that goal is eventually realized across the Indonesian archipelago include effective safeguards to ensure that local forest uses are in fact sustainable. This will be best accomplished through a system of government-led oversight assisted by NGOs and the members of each community.

ESTABLISH EFFECTIVE MECHANISMS FOR INDEPENDENT CITIZEN MONITORING OF TRENDS AND THREATS RELATED TO FOREST LANDS AND RESOURCES.

The current climate of *reformasi* provides a chance to bolster the role of the institutions of civil society—NGOs and community-based groups of forest-resource users—as "watchdogs" over forest policy and practice. But if civil society is to assume an enhanced role in monitoring forestry policies and activities, the capacity of NGOs and local communities to gather and disseminate forest-related information must be strengthened.

It has long been recognized that data and information on forests and forest policies in Indonesia are flawed and incomplete. Field data on forest cover, deforestation, and the impacts of logging, plantations, transmigration, and other activities on forestlands are seriously deficient. Information on the traditional forest management practices of millions of forest dwellers has been sketchy and is often biased toward the interests of industrial forestry projects and investors, with whom local communities are frequently in conflict. Data collected by the Ministry of Forestry and the private sector have long been treated as secret and have been only reluctantly, if at all, shared with Indonesia's citizens. And the long-entrenched bureaucratic culture of *asal bapak senang* ("keep the boss happy") has meant that local forestry officials were reluctant to report poor logging concession performance, illegal logging, or conflicts over resource allocation and use.

Data collected by the Ministry of Forestry and the private sector have long been treated as secret and have been only reluctantly, if at all, shared with Indonesia's citizens.

For their part, NGOs and affected local communities have long tried to document abuses of the law by logging firms and other large-scale forest resource users, but their efforts have been piecemeal and have often been hampered in the field by lack of technical expertise and by opposition from local authorities. NGOs have also tried to document forestry success stories, such as the sustainable local management of damar (*Shorea spp.*) forests in the Krui area of Lampung Province, Sumatra, but have lacked the capacity to do so systematically. The recent effort by a number of NGOs to develop an independent forest development monitoring network— Forest Watch Indonesia—illustrates one strategy for developing citizen monitoring. (*See Box 15.*)

---

15 **FOREST WATCH INDONESIA: AN EXPERIMENT IN CITIZEN MONITORING OF FOREST STATUS AND DEVELOPMENT**

Since late 1997 a number of Indonesian NGOs have been working together to develop Forest Watch Indonesia (FWI), an independent, decentralized early-warning monitoring network for tracking logging, plantation development, mining, and other large-scale development activities within and around Indonesia's major remaining blocks of natural forest.[1] FWI's core task is to gather and analyze information on Indonesia's forestlands and resources and make it available to all interested audiences in a useful and accessible form. Key FWI datasets under development include:

■ baseline data on the status of

Indonesia's forests (type, coverage, condition, infrastructure such as roads, utilization, human settlements, population, and traditional claim areas),
■ existing and planned development projects (logging concessions, industrial timber plantation concessions, estate crop plantations, mining concessions, infrastructure projects, and transmigration project areas),
■ conflicts over forestlands and resources (types of conflicts, parties involved, description, location, time period, etc.),
■ data and analysis covering the economic, political, and legal aspects of forest policy and related conflicts, and

■ documentation of forestry management success stories, including both well-managed logging concessions and local community forest management systems.

Interest and support among NGOs and forest policy reformers within the government for the kind of independent forest monitoring network that FWI is developing is strong. Needed now are working linkages with sympathetic forest policymakers, technical experts, and donor agencies. Of particular importance will be a two-way sharing of data and information between the FWI on the one hand, and, on

the other, official efforts such as those proposed by the World Bank, to map current forest status and monitor field performance of concessions.

Note:

1. Forest Watch Indonesia is the national "node" of Global Forest Watch, an initiative of the World Resources Institute that supports development of a decentralized, independent forest monitoring network spanning the major forest countries of the planet. Telapak Indonesia, a collaborating partner of this report, hosts the secretariat for Forest Watch Indonesia.

The strong citizen participation component in the newly launched timber certification system is another useful example. (*See Box 16.*)

STRENGTHEN AND INTENSIFY THE MULTI-INTEREST DIALOGUE ON FOREST POLICY REFORM THAT BEGAN IN 1998.

Effective and durable forest sector reforms cannot take place without the participation and support of key stakeholder groups, including concerned NGOs, representatives of indigenous and forest-dependent peoples, the various elements of the commercial forest resources sector, the academic community, and representatives of local and national government. This is a simple and obvious point that has been made countless times around the globe— but it is worth noting that there has been no such process in Indonesia in the past 30 years. Rather, the forest policy process has been characterized by:

■ a centralized, pyramidal hierarchy and secretive processes for making decisions about projects and expenditures;
■ strong reliance on traditionally trained professional foresters in top management positions and a corresponding lack of social science expertise and perspectives;
■ a close relationship between the forestry service and the timber industry, amounting in many cases to *de facto* control over policies by major industry players;
■ an urban and upper-middle-class bias among policy-level foresters;
■ a strong colonial forestry tradition and background;
■ patterns of forest sector donor assistance that are technically based and executed in cooperation with the forestry bureaucracy—and that therefore tend to reinforce existing structures and ways of doing things; and
■ the belief that local land use practices are destructive and the resulting assumption of a policing role to limit local access and use.[223]

The dialogue initiated by the Reform Committee established by the Ministry of Forestry and the Community Forestry Forum, discussed above, is a hopeful sign that this long-standing paradigm is beginning to change. These efforts need to receive continuous support from senior government officials, especially in the Forestry Ministry, and Parliament, which is likely to play a much stronger role in policymaking than in the past. These groups should take advantage of the news media —now largely free of the heavy-handed censorship of the Suharto era—to get their ideas and proposals before the public.

Thus far, however, these and other dialogue processes have shown a Jakarta-centered bias. It is crucial that a similar process begin in key forest-resource provinces as well. Provincial governments are much less visible as policymakers than are high-level Jakarta officials, but they are often the ones making the *de facto* decisions about conversion of forestlands to plantations and other uses. And as pressure for decentralization grows, the role of provincial and local governments will increase.

REFORM LOGGING PRACTICES AND BROADEN FOREST UTILIZATION TO INCLUDE MULTIPLE USES AND A WIDER VARIETY OF USERS.

Commercial logging, as currently practiced, causes the degradation and eventual destruction of some 64 million ha of Indonesia's forests—the area allocated for production. From the perspective of preventing fires, logging reform is essential, for two main reasons. First, as was shown in the discussion of the impacts of the 1982–1983 East Kalimantan fires, logged-over forests—particularly if large amounts of slash are left behind— are far more prone to burn than intact forests. Second, when production forest areas are degraded by poor logging practices, they immediately become targets for conversion to plantations and other uses, a process that greatly increases the incentives for, and probabilities of, widespread use of fire to clear land.

Two broad tasks have to be carried out in order to move the current wasteful and inequitable utilization of Indonesia's natural forests toward sustainability and equity. First, the existing system under which timber is produced —both on legal timber concessions and illegally—must be reformed. Second, the framework for natural forest utilization needs to be broadened to encompass an ecosystem perspective and to incorporate a wider range of forest resource uses and users than has been the case. To that end, the steps described below need to be taken over the next several years.

■ CARRY OUT FIELD-LEVEL ASSESSMENTS OF ALL OPERATING LOGGING CONCESSIONS AND REVOKE THE LICENSES OF THOSE THAT HAVE SUBSTANTIALLY VIOLATED THE TERMS OF THEIR CONCESSION AGREEMENTS.

It is widely recognized that a great many of Indonesia's more than 400 logging concessions are violating the terms of their concession agreements.[224] One study in the early 1990s estimated that only 4 percent of concessionaires followed the regulations, and, in 1991, the minister of forestry himself estimated that only some 10 percent of firms obeyed the law. In 1995, the head of East Kalimantan's Forestry Service told a researcher that at least "80 percent [of concession holders] are liars" with respect to their logging practices.[225]

The legal power to revoke logging contracts for violation of their operating terms is clearly spelled out in the 1970 Government Regulation on Forest Utilization and reiterated in the 1999 regulations that replaced it.[226] The main problems in the past have been lack of political will on the part of the industry-dominated Forestry Ministry and lack of capacity to monitor concession performance in the field, a task that is made more difficult by pervasive collusion between logging firms and local officials. The huge economic clout of the industry, worth some $8 billion in exports in 1999 (for timber and pulp together), creates additional disincentives for strong enforcement measures, especially in the current economic climate.

But political will to get tougher on errant logging firms seems to be growing. In September 1998, a top official of the Forestry Ministry testifying before Parliament was urged by legislators to clamp down on violators. He responded that the government had revoked the contracts of at least 86 firms in the previous 10 years (1988–98) and that many more firms had been fined for minor violations. The legislators urged the ministry not only to revoke contracts but also to sue violators for breach of contract, pointing out that "where their concession permits are revoked, it is the government which in fact suffers the losses because it has to manage the damaged forests while the concessionaires have, in a way, benefited from the forests." [227]

The capacity to monitor logging practices must be drastically improved if poorly performing concessionaires are to be identified and eliminated. This is a legal as well as a practical matter: There are numerous cases in which logging firms called to account by the Forestry Ministry have denied the allegations and challenged the ministry to provide hard evidence—something the ministry was frequently unable to do.

The independent Forest Watch Indonesia monitoring initiative (*see Box 15*) is a potentially useful contribution to building capacity to monitor concession performance and violations, but realistically, this effort can at best only complement monitoring by the government itself. To that end, the World Bank has floated a proposal to form three or four forest operation inspection teams led by the Forestry Ministry but possibly including participation by the Indonesian Ecolabeling Institute (LEI) (*discussed in Box 16*) and other institutions such as NGOs and university forestry faculties. Working with an agreed set of criteria and indicators, these teams would carry out field inspections of concession operations.

Given the vast areas to be covered and the large numbers of individual concessions, these teams would have to strategically target their investigations toward concessions with especially bad reputations, areas of high ecological value (such as those bordering important national parks or watersheds), and areas where concession operations are known to be the cause of social conflicts. In making these strategic decisions, the government monitoring apparatus could rely in large part on information supplied by LEI and by NGO initiatives such as Forest Watch Indonesia.

■ **CHANGE THE ECONOMIC INCENTIVES THAT ENCOURAGE WASTE AND A "CUT-AND-RUN" MENTALITY BY LOGGING CONCESSIONAIRES.**

Many of the timber industry's wasteful practices, including the reckless use of fire, stem from the distorted economic incentives arising from the current concession system. Reform of the concession system and its taxation and pricing mechanisms is a complex matter that has been dealt with extensively elsewhere.[228] Four basic measures, however, seem to be essential prerequisites for improving concession performance.

Allocate concessions by auction. This measure, also required by the IMF agreement, will help give concessionaires incentives to maintain the value of forest resources under their stewardship, since they will have paid a substantial price to obtain the concession and will have the right to auction it off in the future. Originally, the IMF agreement required that such a system be in place by the end of June 1998. This was an unrealistic goal and was not met, although auction options are under active discussion in the Forestry Ministry. The idea is good, but rushing to implement it without adequate discussion among stakeholders and without a period of experimentation may create more problems than it solves. There must also be a clear division between those forest areas available for auction on the free market and those reserved for community forest concessions. Local communities—even with backing from donors and international NGOs—will never be able to match bids from the private sector for access to truly valuable forestlands, and a policy that de facto restricts local communities to degraded forestlands would work against the objective of sharing greater forest-based economic benefits with local and indigenous communities.

TIMBER CERTIFICATION IN INDONESIA:
THE INDONESIAN ECOLABELING INSTITUTE (LEI)

The Indonesian timber industry has been overtly concerned with timber certification since at least 1990. In that year, Indonesia hosted a meeting of the International Tropical Timber Organization (ITTO) at which member states pledged that by 2000 all tropical timber in international trade would come from sustainably managed sources.

There was considerable concern in Indonesia during the early 1990s that bans and boycotts on tropical timber in importing countries might mushroom into a significant problem for the industry, and in 1993 the Indonesian Forest Industries Association (APHI) formed a team of experts to develop principles and criteria for implementing the ITTO target in Indonesia. At the end of that year, the minister of forestry supported establishment of the Indonesian Ecolabeling Working Group, headed by former environment minister Emil Salim. This group worked quietly for four years to develop and test processes, criteria, and indicators for an Indonesian timber certification system. It brought in representatives from the APHI team of experts, the Forestry Ministry, the National Standards Agency, major forestry faculties, and NGOs.

Beginning in 1996, the working group tested its procedures, criteria, and indicators in the field with 11 logging concessions that agreed to serve as experimental subjects. In February 1998, the Indonesian Ecolabeling Institute (LEI) was formally constituted as a legal body under the auspices of the LEI Foundation, the board of which defines LEI policies. A detailed series of process, criteria, and indicators documents were finalized soon thereafter in multistakeholder workshops and were officially adopted by the National Standards Agency in June 1998 as the Sustainable Forest Management Certification System for Production Forests.

LEI was officially launched in September 1998. Since that time, five firms have sought certification. One operation in Sumatra has been certified, three were in process as of March 1999, and one in East Kalimantan failed due to forest fires in the concession during early 1998.

The LEI system is voluntary at present, and it applies to specific forest management units, not entire companies. Procedures, criteria, and indicators for all steps in the process are spelled out in the LEI documents mentioned above. There are essentially four stages to the process: preliminary evaluation by an expert panel; field assessments by assessors certified by LEI, with comments by local stakeholders invited; performance evaluation by an expanded expert panel; and certification, good for five years. (*See Appendix C.*)

At this early stage (the first official certification processes began in mid-1998), LEI is not only acting as a certifying body, it is also conducting training programs for assessors and expert panel members, and carrying out some assessments in the interim. LEI's objective over the next several years is to leave both assessor training and field assessment to other educational institutions and certifying bodies and restrict itself to accrediting and monitoring the certifying bodies.

Why should Indonesian logging firms want to seek voluntary certification? Given the main markets for Indonesia's wood—domestic for most sawnwood and furniture, East Asian for 43 percent of all exports (mostly plywood)—it seems unlikely that producers (aside from the small number that can cultivate special niche markets in western countries) will receive an appreciable "environmental premium" for certified timber.[1] Regulatory relief is a more persuasive incentive; 137 separate regulations apply to logging concessions, and a recent ADB study estimated the annual costs of

Indonesia's onerous and complex regulatory regime at about $98 million.[2] If certification meant official regulatory relief, the costs of the certification process might look extremely attractive to logging firms. One proposal currently under discussion would link the degree of regulatory relief granted to a company to the grade it received in its certification assessment.

Additional incentives arise from the pressures for reform shaking Indonesia. It is becoming more and more difficult for loggers to conduct "business as usual" in the face of popular sentiment for fundamental changes in forest policy and practice. Certified firms are less likely to find themselves in the firing line of populist anger and policy reforms.

The greatest challenges facing certification are to build capacity for assessing and monitoring logging management units and to create an efficient and speedy bureaucracy to administer the certification system. If the supply of efficient, honest, and highly regarded certification services cannot keep up with demand, the system will wither away. This is an area in which international aid agencies concerned about the future of Indonesia's forests should provide strong financial and technical support.

Finally, the LEI system needs to secure international recognition. Many legitimate and less legitimate certification schemes have sprung up in the past decade, and international buyers are increasingly confused. Allaying this confusion by providing a "one-stop shop" for international certification of certifying bodies is one of the primary missions of the Forestry Stewardship Council (FSC). But the idea of an international body "certifying" the Indonesian system—and therefore being somehow superior to it—is politically unacceptable to the logging industry and to many other stakeholders within Indonesia. In early 1998, LEI and the FSC signed an

agreement to resolve this issue and find ways to harmonize the two systems, and joint technical cooperation and international observation began in November 1998. LEI, for its part, wishes to achieve formal FSC recognition of the Indonesian system and it believes that opening the Indonesian system to international scrutiny will increase its credibility. Given the close substantive match between the principles and criteria of LEI and the FSC, and Indonesia's status as one of the world's most important timber producers, it seems likely that an accommodation will be reached, especially since Malaysia, another major producer, has recently adopted the same position vis-à-vis "mutual recognition" with the FSC.

In the longer term, LEI does not plan to restrict itself to certifying logging operations. Nontimber forest products, timber and other plantations, and marine products are all areas for which certification has been discussed.

Sources:

LEI, 1998a, 1998b; ADB, 1997.

Notes:

1. Forty-three percent of Indonesia's wood product exports goes to East Asian markets, mostly in the form of plywood; 26 percent goes to Europe and North America, mostly as finished products and molding and other wood-working products. The remainder goes to other Southeast Asian countries, the Middle East, and Africa. The market for hardwood plywood is rising in North America, however, in line with the use of panel products in building materials (personal communication, Mubariq Ahmad, Executive Director, Indonesian Ecolabeling Institute, March 28, 1999).

2. ADB, 1997.

Institute area-based government charges for timber extraction. Currently, concessionaires pay volume-based charges at set rates for the timber they extract. This has two primary negative effects. First, it requires the government to monitor the flow of logs in order to assess charges, or, as is actually done, to rely on the estimates provided by concessionaires. The government's lack of capacity to make its own accurate assessments leads to cheating and to diminished government revenues. Second, since loggers are taxed on what they ship out rather than on the volume of trees they cut or otherwise damage, there are no incentives for avoiding waste. As a result, according to the World Bank, each year some 8 million m³ of timber are left by loggers to rot in the forest.[229]

This destructive incentive system should be replaced with an area-based charge levied on a per-hectare basis for the whole of a concession. Because timber stand inventories for most concessions are incomplete or inaccurate, it is impossible, for the time being, to set variable rates on the basis of local conditions. A recent report by the Indonesia-U.K. Tropical Forest Management Programme argues that area fees should initially be set at $20 per hectare, well below the available economic rent from the most productive concessions but probably close to total rent for less productive ones. In the longer term, once more detailed stand inventories are available, the tax liability of each concession unit can be based on local conditions.[230] Political support for this reform comes from the IMF program, which calls for development of a new forestry "resource rent tax" of some kind.

Introduce performance bonding on concession operations. Performance bonds, which are also mandated by the IMF program, would give the government additional leverage over concessionaires to ensure that their operations are well managed. The World Bank suggests that 30 percent of average annual operating costs be posted as bond or that 40 percent be collected as a deposit on total resource rents payable. Interest would be returned to the operator, but the operator would be obliged to immediately replenish the bond if it is collected or drawn down by the government for any breach of concession conditions.[231]

Delink logging from processing industries. During the 1980s, the government, in an effort to boost value-added processing, primarily of plywood, obliged logging concessionaires to establish their own processing ventures. The result of this policy was to create combined logging-and-processing firms with a single-minded devotion to acquiring enough raw material to keep their often inefficient processing facilities working at a profitable proportion of capacity. As timber shortages have developed,[232] this arrangement has boosted incentives for firms both to overcut their own concessions and to acquire illegally logged timber.

Delinking would allow logging firms to sell their output to the highest bidder, rather than subsidize their own processing plants, and would be likely to drive the more inefficient processing facilities out of business. That, in turn, would reduce demand for timber and decrease the pressures for overcutting and illegal logging.

■ **ACCELERATE THE EVOLUTION OF THE CURRENT CONCESSION SYSTEM TOWARD PERMANENT FOREST MANAGEMENT UNITS.**

Since 1991, the Ministry of Forestry, with assistance from the United Kingdom's aid program, has been developing a wholly new model for organizing timber production at the field level. Essentially, this effort seeks to redraw current concession boundaries to reflect the actual state of the forest resource, nonforest land uses, and local community territories and needs, and to use this information to establish permanent forest management units (*Kesatuan Pengusahaan Hutan Produksi,* or KPHPs). The KPHPs would form the basis for allocating new concessions—and reallocating existing concessions, where necessary—in areas large enough to be economically viable but within boundaries drawn through a participatory process. This system, it is hoped, will minimize the pervasive conflicts with local communities and the overlaps with other forest and development activities (such as plantations) that have long characterized the sector. Five pilot KPHPs had been established by late 1996, and the government appears committed to continuing the development of this approach.[233]

The KPHP model provides the basis for bringing the macro-scale mapping of the country's forest resource base and condition, recommended above, to the ground level. It also provides a potential vehicle for incorporating local community interests into decision-making about how the forest resources of a KPHP unit are to be utilized. Once the legal basis for community concessions is in place, these legal understandings could and should be incorporated into the KPHP system.

RETHINK AND REFORM THE PLANTATION SECTOR

The fast-growing pulp and paper and oil palm plantation sectors are exerting heavy pressure on the forest and were the main culprits behind the 1997–98 forest fire disaster. The clear delineation of the nation's permanent forest estate and the establishment of unambiguous legal protection against its conversion, as recommended above, are the most important first steps in restraining the destructive role that plantation development is currently playing. Key additional actions that need to be taken include the following.

■ **INSTITUTE A MORATORIUM ON GRANTING NEW CONCESSIONS FOR OIL PALM, TIMBER, AND OTHER PLANTATIONS UNTIL A NATIONAL INVENTORY OF PERMANENT FOREST ESTATE IS COMPLETED.**

Freezing the allocation of new lands for plantation development will no doubt be difficult politically, given the current economic crisis and the government's desire to boost exports of pulp and oil palm. Nevertheless, the only honest and practical way to ensure that lands not suited for plantation development (forested lands, areas used by local communities, and nonforest lands better suited for other purposes) are not misallocated is to wait until at least a preliminary inventory of forestlands has been carried out.

One approach that might ameliorate the inevitable industry and political opposition to such a move would be to prioritize for early inventory provinces, and areas within provinces, slated for plantation development. The moratorium could then be lifted area by area, once an inventory is conducted and areas of permanent forest estate and local use are excluded from consideration.

■ **BAN THE ESTABLISHMENT OF PLANTATIONS ON ALL BUT TRULY DEGRADED FORESTLANDS.**

This is, in theory, already government policy, but it has been widely flouted by both timber and oil palm plantations.[234] Key to strengthening implementation of this policy is a restriction on the power of governors to allocate lands for plantation development. In their haste to promote economic growth and increase their own wealth, provincial governors have frequently allocated permanent production forest areas, and even protection forests, for plantation development. The role of provincial governments in drawing up provincial land use plans and allocating plantation concessions needs to be overhauled. Frequently, the concession is granted and then the plan is changed accordingly.[235] Until reforms that make governors more accountable to the Ministry of Forestry—and to their own citizens—are put in place, their power to make land-use plans and allocate land for development should be substantially restricted by the central government.

> *The fast-growing pulp and paper and oil palm plantation sectors are exerting heavy pressure on the forest and were the main culprits behind the 1997–98 forest fire disaster.*

■ **REVISE THE INCENTIVE STRUCTURE FOR TIMBER PLANTATIONS SO THAT COMPANIES ARE NO LONGER ENCOURAGED TO CUT NATURAL FOREST.**

The perverse system of incentives that encourages pulp and paper firms to rely on natural feedstock rather than on their own plantation wood has been well documented.[236] To change this, the following steps should be undertaken:

■ Deny timber plantation firms access to large areas of forested lands as a source of feedstock.
■ Utilize the community concession model recommended above to transfer management (or shared management) of truly degraded lands to local communities that would grow pulpwood and supply it to existing mills as part of their concession agreement.
■ For firms currently holding valid concession contracts, establish "sunset clauses" setting specific dates beyond which they may no longer use natural forest to supply their mills.[237]

Further into the future, completion of a reliable forest (and degraded forestland) inventory will allow siting of pulp mills in areas distant enough from permanent forest areas that incentives for using natural forest feedstock will decrease, and monitoring possible illegal transport of natural forest timber to the mill site will be easier.

■ **SAFEGUARD THE INTERESTS AND LIVELIHOODS OF LOCAL COMMUNITIES IN PLANTATION AREAS.**

The recent boom in timber and oil palm plantations has caused some of the worst conflicts with local communities in forest areas. To reduce the level of local conflict and the great economic losses that local communities have been forced to bear, the government should establish meaningful "social acceptability" criteria and mechanisms to ensure that plantation development is contingent on the full and informed consent of local communities and provides benefits to the community equal to or greater than those they are obtaining from their existing access to forest resources. The principle is that local communities must not be made worse off by plantation establishment. The determination should be made by the communities themselves, not by actors (the company and the local government) with a vested interest in seeing the plantation go forward.

■ **STRENGTHEN RULES AND PENALTIES AGAINST CLEARING PLANTATIONS WITH FIRE.**

When government satellite analyses in September 1997 indicated that plantations were responsible for a great proportion of the intentionally set fires, the government made a big show of threatening to revoke their operating permits. In the end, this came more or less to nothing, since any company that provided an alibi by a certain date—hard evidence not being required—was absolved of responsibility. Rules are already on the books limiting the use of fire for plantation development, but they are not strict enough and, judging by the experience of 1997, are not much enforced.

Obtaining evidence strong enough to withstand a legal challenge from a company accused of burning is a significant obstacle for the government, since its capacity to monitor forestry activities in the field is limited. Three steps would help remedy this situation:

■ The development of government and independent citizen monitoring capacities would provide a much stronger factual basis for holding corporate arsonists to account.

■ A regulatory change in the burden of proof in favor of the government would help: where remote-sensing data provided *prima facie* evidence that a company was using fire to clear land, the company would be presumed guilty unless it could produce sufficient proof that it had not intentionally set fires. (The standard for such proof would have to be defined.)

■ Requiring companies to post a performance bond similar to the one recommended for logging concessions would put some teeth into government enforcement efforts and provide a financial incentive for compliance.

DECLARE A FIVE-YEAR MORATORIUM ON THE TRANSMIGRATION PROGRAM AND RECONSIDER ITS OBJECTIVES AND METHODS

The transmigration program has long been bedeviled by many of the same deficiencies found in the plantation sector. Incomplete and inaccurate data on the physical and socioeconomic characteristics of prospective sites, combined with the government's casual attitude toward converting forest and displacing local communities, has led to one disaster after another as sites were developed on inappropriate soils and in areas already being used by local people. Poor site preparation methods, such as the complete removal of tree cover and the extensive compaction of soil, have characterized the program.

It would therefore be wise to declare a five-year moratorium on the establishment of new transmigration sites to allow time for completion of the national inventory of forest areas and degraded forestlands. With that information in hand, new transmigration sites—should the government decide to continue the program at all—could be located in areas that are truly degraded, are unencumbered by private ownership rights, and pose no substantial risks of conflict with local land uses and resource needs.

Both transmigration and plantation development should be limited to sites within that category and might be combined in joint initiatives, as is already the case. Indeed, very little land meets the tests of being suitable for rice or other annual crop agriculture, not covered by forest that should be part of the permanent forest, and not encumbered by legal or customary ownership rights. Accordingly, any future transmigration programs should be linked to cultivation of tree crops such as fast-growing pulp species or oil palm rather than to rice cultivation.

SUMMING UP

This report has intentionally refrained from offering detailed recommendations for restructuring the institutions that either formally control forest policy (principally, the Ministry of Forestry) or play an important direct or indirect role in determining the fate of Indonesia's forests (such as provincial governors and officials in the transmigration, mining, and infrastructure sectors). The institutional questions will need to be worked out incrementally on the basis of needs that arise, institutional deficiencies that are identified, and political compromises that must be struck in the process of formulating and implementing the forest policy reform agenda.

Three principles related to governance and institutions, however, should animate the process of institutional restructuring and renewal that must accompany forest policy reform.

First is the crucial importance of developing transparent, multi-stakeholder processes—both institutional and political—for debating, deciding on, and implementing policy and institutional reforms. Freedom of information for the public, and effective ways of holding political leaders and policymakers accountable for their decisions (such as free and fair elections and meaningful legal remedies with which citizens can challenge and overturn bureaucratic decisions), are essential elements of such a process. To readers not familiar with Indonesia's recent political history, this may sound like a rather unexceptional point. But for more than three decades, since the exploitation of Indonesia's forests began in earnest, key policy decisions related to forests (and almost everything else, for that matter) have been tightly controlled by small groups of unelected officials and their patrons in the large cartels that have dominated the economy.[238] Changing this political system and culture, together with recovery from the economic crisis, will be Indonesia's major preoccupations during the first decade of political reconstruction for the 21st Century. Questions related to forest policy will by no means dominate that process. But with three-fourths of the country designated as "forestland" and with the increase in violent conflicts over control of those lands and their resources, debates over forest policy will certainly be an important political issue.

Second, the issue of decentralization of governance must be treated very carefully or it may accelerate the rate of forest degradation and increase the level of social conflict over forest resources. There is some tendency to equate *reformasi* with decentralization, since the tendency of the Suharto regime was to centralize power and decisionmaking as much as possible. But without significant overhaul of the provincial and local systems of government, unthinking decentralization of land and forest management powers to provincial and sub-provincial levels is likely to result in an explosion of conversion of forests to plantations and other nonforest uses. In the case of East Kalimantan, the local government "seems to be using reform sentiment to boost its power and speed conversion. . . .Turning reform sentiment for greater regional autonomy to his advantage, the governor said, 'In the spirit of reform, the [central] government is expected to issue a ruling which will allow the local administrations to do all the licensing work [for new oil palm plantations].'"[239]

Before decentralization of decisionmaking over forest lands and resources is allowed to proceed, the principles of accountable and transparent governance and multi-stakeholder decisionmaking processes that were discussed above must be put into practice at the provincial and sub-provincial levels. In addition, such powers should not be given to provincial officials until the national inventory of permanent forest estate and lands available for conversion is complete. Within that framework, accountable provincial governments should certainly be granted a share of power in deciding what investments (a transmigration site versus a timber plantation, for example) should be made on specific pieces of land that have been determined by the national inventory to be available for nonforest uses. But delineation of the nation's permanent forest estate is a national matter and should not be influenced or controlled by the vagaries of local politics and vested interests.

Finally, the restructuring of the Ministry of Forestry—which seems very likely to happen over the next several years—will be a complex and drawn-out process. As the World Bank points out, "Ministries of government—particularly large and powerful ones like the Ministry of Forestry—cannot be reformed—or replaced—in a short time frame."[240] But the reform processes have to be implemented by the existing bureaucracy; they cannot wait for the bureaucracy to be restructured. Serious thought needs to be given to an interim institutional arrangement that taps the considerable expertise and experience in the forestry bureaucracy but at the same time provides oversight and enforcement to ensure that reactionary elements within that bureaucracy do not slow or sabotage reforms embodied in the new laws and policies that have been established or will soon be in place.

It is important to remember that institutions are more than just the sum total of their organization charts and the regulations that they live by and implement. Anyone who has worked closely with the Ministry of Forestry on a day-to-day basis (as have the authors of this report) would agree that many officials are dedicated professionals who have long been disillusioned with the policies they were obliged to implement and the behind-the-scenes collusion and corruption that they saw but could not do anything about. If Indonesia is to embark upon this new millennium with forest policies that ensure a more sustainable future for the nation's forests and a more prosperous life for the country's millions of forest-dependent people, the forces of reform that are storming the walls of the forestry bureaucracy must make common cause with those inside whose hearts lie with the reformers and whose skills and experience are so important for the future of the reforms that so many Indonesians—and their friends around the world—fervently hope will come to pass.

ABOUT THE AUTHORS

Charles Victor Barber is a Senior Associate in WRI's Biological Resources Program. He has been with WRI since 1989; since 1994, he has been based in the Philippines. Dr. Barber is a specialist on Southeast Asia and has worked extensively on Indonesian forestry policy, conservation of marine biodiversity, and biodiversity policy. He has written numerous publications on Indonesian forestry policy and political economy, strategies for combating destructive fishing practices on coral reefs, protected areas management, the Convention on Biological Diversity, and on issues relating to bioprospecting and access to genetic resources. Prior to joining WRI Dr. Barber lived in Indonesia for three years, working as a consultant on Indonesian forestry and environmental issues for a variety of international donor agencies. He received a Ph.D. in Jurisprudence and Social Policy, law degree, and M.A. in Asian Studies from the University of California, Berkeley.

James Schweithelm, a geographer by training, first visited Indonesia as a doctoral researcher in 1985. His research focused on watershed management in South Kalimantan province in cooperation with the Indonesian Ministry of Forestry and the East-West Center. Since then, he has spent eight years in various parts of Indonesia working on a wide range of issues related to forest conservation and management. He was the Forest Policy Officer of the World Wide Fund for Nature Indonesia Programme during the 1997-1998 forest fire episode, putting him in a position to closely observe the tragedy as it unfolded. He is familiar with forest issues throughout the Asia-Pacific region and worked on forest policy in Nepal for two years. He is currently a freelance consultant living with his family in Burlington, Vermont.

116 Byron and Shepard, 1998; Vayda, 1998.

117 Tomich and others, 1998. On the historical use of fire as a weapon of resistance in the forests of colonial Java, see Peluso, 1992.

118 World Bank, 1999c.

119 Personal communication, GTZ IFFM/SFMP, Samarinda, Indonesia, October 21,1999.

120 GOI and IIED, 1985.

121 Romm, 1980.

122 Gillis, 1988.

123 Mackie, 1984.

124 Gellert, 1998.

125 Kartidihardjo and Supriono, 1999.

126 Barber, Johnson, and Hafild, 1994.

127 World Bank, 1995.

128 Barber, Johnson, and Hafild, 1994.

129 "Wood-Processing Firms to Face Log Scarcity: Minister," *Jakarta Post*, October 1, 1998.

130 World Bank, 1993.

131 Ibid.

132 Indonesia–U.K. Tropical Forest Management Programme, 1999.

133 "Timber Fencing and Smuggling Still Rampant"; "Legislators Urge Government to Stop Timber Brokers." *Jakarta Post*, July 3, 1996.

134 EIA and Telapak Indonesia, 1999.

135 Sunderlin, 1998.

136 Ibid.

137 The 1991 Indonesian Forestry Action Programme stated that "the role of plantation forests in supplementing natural forest resources will also be very important to conservation objectives in the country " (GOI 1991, vol. 2: 60).

138 The main subsidy—apart from almost free access to land—was a seven-year, interest-free loan covering 32.5 percent of plantation establishment costs, drawn from the Reforestation Fund (Potter and Lee, 1998b).

139 World Bank, 1999c.

140 Kartidihardjo and Supriono, 1999.

141 "Wood-Processing Firms to Face Log Scarcity: Minister," *Jakarta Post*, October 1, 1998.

142 World Bank, 1998b.

143 "Looking Ahead to the Next Century," *Paper Asia*, February 1997, pp. 7–10.

144 "Indonesia Planning a Further Mass Pulp Capacity," *PPI This Week*, July 29, 1996.

145 "Indonesia Prepares to Tap Plantations," International Woodfiber Report, June 1997.

146 World Bank, 1993.

147 Ibid.

148 Barber, 1997.

149 "BUMN, Grup Besar Ikut Bakar Hutan" [State firms, conglomerates, burned the forest], Media, September 18, 1997.

150 State Ministry for Environment, 1998: xi.

151 CIC Consulting Group, 1997

152 World Bank, 1999c.

153 CIC Consulting Group, 1997

154 Kartidihardjo and Supriono, 1999.

155 Ibid.

156 Potter and Lee, 1998b.

157 Kartidihardjo and Supriono, 1999.

158 "Forest Fires Mostly in 'Plantation Areas,'" *Jakarta Post*, October 9, 1997.

159 Wakker, 1998.

160 "Forest Fires Mostly in 'Plantation Areas,'" *Jakarta Post*, October 9, 1997.

161 *Reuters*, April 3, 1998.

162 Wakker, 1998.

163 GOI (Government of Indonesia), 1993. Rencana Pembangunan Lima Tahun Keenam (Sixth Five-Year Development Plan). Jakarta.

164 World Bank, 1994b.

165 Potter and Lee, 1998b.

166 CIC Consulting Group, 1997.

167 Maltby, 1997.

168 See, for example, Rieley and Page, 1997; Maltby, Immirzi, and Stafford, 1996.

169 Tim Teknis Pengembangan Lahan Gambut Di Propinsi Kalimantan Tengah [Technical Team for Peatland Development in Central Kalimantan Province], 1997.

170 Boehm, 1999.

171 The initial EIA was carried out by a team from the respected Bogor Agricultural Institute, covering the first-phase area for the project, approximately 227,000 ha. While extremely detailed, the study adopted the curious device of balancing the serious negative environmental impacts it predicted against the "positive impacts" of implementing a policy that the government wanted implemented. Nevertheless, this study was not positive enough for the Public Works Ministry, which produced its own EIA for the entire project in early 1997. *(See note 169.)* This study came to the conclusion that while

> In a quantitative sense [the project] has more important negative impacts than positive impacts....the negative impacts are compensated for by the positive impacts. This is because the positive impacts, although few in number, are of a strategic character for both the region and the nation. These include regional economic development, equalizing development among regions, and reducing the burden on heavily populated areas such as Java and Bali..."

In short—and in clear violation of Indonesia's environmental impact laws—the important negative environmental impacts were discounted because they stood in the way of something that the president wanted to do.

172 "The Mega-Rice Project, Central Kalimantan, Indonesia: An Appeal for Intervention to the International Community," 1998. Briefing dossier compiled by SKEPHI Support Office in Europe. Amsterdam.

173 Vidal, 1997.

174 In what may have been a form of silent protest, the Environmental Management Agency (BAPPEDAL) main office in Jakarta put on the wall of its reception area a large Landsat photo from early September of the PLG project burning and emitting huge quantities of smoke.

175 Vidal, 1997. One of the authors (C. V. Barber) accompanied Vidal on his survey of the PLG project area in October 1997 and returned to the area in November 1997.

176 The amount was Rp 1.1 trillion, equivalent to about $350 million at the mid-1997 exchange rate of approximately Rp 3,000/$1. "Rp. 1.1 Triliun Habis di Gambut" [Rp. 1.1 trillion spent in the peat swamp], Kontan, October 13, 1997, Jakarta.

177 Ibid.

178 "Di PLG Kapuas, Tak Ada Ganti Rugi" [In the Kapuas PLG Project there is no compensation], *Palangkaraya Post*, September 3, 1997.

179 "Optimis, Oktober Ini Panen di PLG" [Optimistic, there will be a harvest this October at the PLG Project], *Palangkaraya Post*, September 5, 1997.

180 "Akibat Kemerau, Penanaman di PLG tak Penuhi Target" [Because of the drought, the harvest at the peat swamp project will not reach its target], *Banjarmasin Post*, August 28, 1997.

181 "Warga Berharap Presiden Panen Perdana di Dadahup" [Residents hope the president will make the first harvest at Dadahup], *Banjarmasin Post*, September 29, 1997.

182 "Bara Api Terus Terlihat di PPLG Sejuta Hectare" [Fires continuously reported at the million-hectare peat swamp project], Kompas, October 1, 1997.

183 WALHI, 1999.

184 "Di PLG Kapuas, Tak Ada Ganti Rugi" ["In the Kapuas PLG Project there is no compensation"], *Palangkaraya Post*, September 3, 1997.

185 "Notulen Rapat Tentang Pemantapan Pelaksanaan Program Lahan Gambut 1 Juta Hectare di Kabupaten Dati II Kapuas" [Minutes of a meeting concerning consolidation of implementation for the 1 million hectare peat swamp project in Kapuas District], April 24, 1996. This agreement narrowly defined the compensation to be paid as applying only to standing crops, trees, fishponds, and houses destroyed in the construction of canals and set very low compensation rates. At current exchange rates, for example, the valuable rattan gardens and beje fishponds were only valued at 5 cents per m³. Even this minimal compensation was never paid, however.

186 Herman, 1998.

187 The elevation of the peatland rises gradually from the Java Sea to the north end of the PLG project area by approximately 12 meters, meaning that the canals essentially create paths for water from the peatlands to drain into the sea. In addition, water levels in the area's major rivers vary greatly, depending on precipitation and other factors, further accelerating the drainage effect (Boehm, 1999).

188 Rieley, 1999.

189 "Government Proposes Resurvey of Million-Hectare Peatland Project," *Antara News Agency*, June 24, 1998.

190 "Bank Report Exposes Chaos at Central Kalimantan Mega-project," *Down to Earth*, No. 39, November 1998, London. With respect to the provincial government's claim that it had nothing to do with the project, it is ironic to note that the governor's son is widely reported to control one of the major contracting firms that cut the canals (Vidal, 1997).

191 "Diteliti, Proyek Transmigrasi Yang Ditelantarkan Perusahaan Inti" [Transmigration projects abandoned by their core companies investigated], Suara Pembaruan, September 16, 1998. The minister's support for turning the PLG area into a giant oil palm plantation accords with the statement (made confidentially to one of the authors) by a member of the PLG Environmental Impact Assessment team in October 1997 that the whole rice-growing justification for the project was a sham and that the real purpose was to establish the canal and road infrastructure necessary to attract oil palm investors.

192 Rieley, 1999.

193 "Walhi sues government over peat land." *Jakarta Post*, September 23, 1999.

194 "Smoke from Indonesian Fires Begins to Cast Pall over Sumatra, Borneo," AFP, August 3, 1999.

195 The grim fate of Dadahup village on the Mengkatip River was featured in *The Guardian*, on CNN, and on BBC Television. This was in part because the traditional leaders in Dadahup were in contact with Jakarta-based NGOs, which steered journalists to their village. Of the Dayak villages affected by the million-hectare megaproject, Dadahup is most easily accessible from the outside world.

196 Vidal, 1997.

197 Goldammer, 1998.

198 Brookfield, Potter, and Byron, 1995.

199 World Bank, 1998b.

200 Ibid.

201 Ibid.

202 P. Waldman, "Desperate Indonesians Attack Nation's Endangered Species," *Asian Wall Street Journal*, October 27, 1998.

203 For a detailed discussion of the perversion of the concept of "the rule of law" in New Order Indonesia, see Barber 1997: 49–53.

204 Peluso, 1992: 93–94.

205 "One Dead as Police Search for Teak-Wood Pilagers Turns Ugly," AFP, September 11, 1998.

206 "Lawlessness Spreads as Looters Defy Army," *Straits Times*, July 20, 1998.

207 K. Yamin, "Politics-Indonesia: 'People Power' to Some, Mob Rule to Others," Interpress Service (IPS), August 18, 1998.

208 Komite *Reformasi* Pembangunan Kehutanan dan Perkebunan [Forestry and Estate Crops Development Reform Committee], Memorandum No. 1, 1998, Jakarta. The committee's terms of reference are included in this memorandum as Appendix III.

209 Kartodihardjo, 1999.

210 Unofficial translation of Statement from the Communication Forum for Community Forestry (FKKM) Meeting, June 22–23, 1998, Jogjakarta, Indonesia, from World Bank, 1998b.

211 "Brunei Threatens to Sue Jakarta if Fires Not Contained," Television Corporation of Singapore, July 29, 1999.

212 "Suharto's Son Calls Accord with IMF 'Neocolonialism,'" Kyodo News Service, January 15, 1998.

213 Kartodihardjo, 1999.

214 Kartodihardjo, 1999.

215 "Revise Forestry Bill, Say Former Ministers," *Jakarta Post*, June 16, 1999.

216 "Indonesia's New Law Still Exploits Forests and Land Rights," *Straits Times*, July 23, 1999.

217 "Indonesia's new president wants federalism." AFP, October 24,1999.

218 World Bank, 1999a.

219 Tantra, Basuki, and Dwiyono, 1998.

220 For analysis of Suharto-era policies towards indigenous and other forest-dependent communities, see Dove, 1988; Zerner, 1992; Guiness, 1994; and Evers, 1995.

221 World Bank, 1998b.

222 Ibid. For a detailed analysis of the troubled history of Papua New Guinea's traditional landowners—who control almost all of the country's forests—and forest exploitation, see Filer and Sekhran, 1998.

223 Barber, 1997: 52.

224 In September 1998 the Ministry of Forestry reported to Parliament that there were currently some 421 private timber companies with concessions totaling 51.5 million ha, while 6 state-owned forestry firms conducted logging on another 4.9 million ha. "Ministry Revoked 86 Timber Permits over Last 10 Years," *Jakarta Post*, September 25, 1998.

225 For a detailed review and analysis of timber concession noncompliance with applicable regulations, see Gellert, 1998: 210–14.

226 Government Regulation No. 6/1999 Concerning Forest Industries and Harvest of Forest Products from Production Forests. The regulation provides sanction for a wide range of violations, including complete revocation of concession rights, decrease in the size of a concession, and monetary fines. In a significant departure from past practice, concession-holders may be sanctioned for failing to "build the capacity and include the participation of local communities adjacent to and in the forest in concession activities" (Article 34(1)(g)).

227 "Ministry Revoked 86 Timber Permits over Last 10 Years," *Jakarta Post*, September 25, 1998.

228 See, for example: Gillis, 1988; World Bank, 1993; Barber, Johnson and Hafild, 1994; World Bank, 1995; Gellert, 1998; World Bank, 1998b. Indonesia-UK Tropical Forest Management Program, 1999.

229 World Bank, 1994a.

230 Senior Management Advisory Team, Indonesia–U.K. Tropical Forest Management Programme, 1998.

231 World Bank, 1998b.

232 In October 1998, the minister of forestry predicted that the country's wood-processing industry would face an annual log shortage of at least 25 million m³, or 45 percent of demand, over the following five years. The IMF-mandated expansion of log exports (via reductions on export taxes), with an annual quota of 5 million m³ in the first year, was a major factor behind this situation, according to the minister. The supply shortage, he noted, "has resulted in rampant wood stealing and illegal trade." "Wood Processing Firms to Face Scarcity: Minister," *Jakarta Post*, October 1, 1998.

233 Natural Resources Institute, 1996.

234 Potter and Lee, 1998b.

235 Potter and Lee, 1998a.

236 For a detailed analysis of the destructive impacts of the current timber plantation incentive system, see World Bank, 1993.

237 These recommendations are drawn from World Bank, 1998b.

238 For a detailed analysis of the closed and nondemocratic nature of policymaking in Suharto-era Indonesia, see Schwarz, 1994; Mackie and MacIntyre, 1994.

239 Potter and Lee, 1998b.

240 World Bank, 1998b.

REFERENCES

In addition to the sources listed below, the following newspapers, magazines, and news services were used:

Agence France Press (AFP)
The Age
Antara News Agency
Asian Wall Street Journal
Asiaweek
Associated Press
Bangkok Post
Banjarmasin Post
Cable News Network (CNN)
Down to Earth
The Economist
Far Eastern Economic Review
The Guardian
Indonesian Observer
International Herald Tribune
International Woodfiber Report
Interpress Service (IPS)
Jakarta Post
Kompas
Kompas Online
Kontan
Kyodo News Service
Media
New York Review of Books
Palangkaraya Post
Paper Asia
PPI This Week
Reuters
Republika
Suara Pembaruan
South China Morning Post
Straits Times
Sydney Morning Herald
Television Corporation of Singapore

ADB (Asian Development Bank). 1997. *Strategy for the Use of Market-Based Instruments in Indonesia's Environmental Management*. Manila: Environment Division, Office of Environment and Social Development.

Allen, B., H. Brookfield, and Y. Byron, 1989. "Frost and Drought through Space and Time. Part II: The Written, Oral and Proxy Records and Their Meaning." *Mountain Research and Development* 9: 279–305.

ASEAN (Association of Southeast Asian Nations). 1997. "Regional Haze Action Plan." Annex to the joint press statement, ASEAN Ministerial Meeting on Haze, Singapore, December 22–23.

BAPPENAS (National Development Planning Agency). 1993. *Biodiversity Action Plan for Indonesia*. Jakarta.

BAPPENAS (National Development Planning Agency). 1999. *Final Report, Annex I: Causes, Extent, Impact and Costs of 1997/98 Fires and Drought*. Asian Development Bank Technical Assistance Grant TA 2999-INO, Planning for Fire Prevention and Drought Management Project. Jakarta. April.

Barber, C. V. 1995. "Integrating Conservation and Development in the Asia Pacific Region: Projects, Policies, Problems and Potentials." In Asian Development Bank and World Conservation Union (IUCN), *Biodiversity Conservation in the Asia Pacific Region*. Manila.

Barber, C. V. 1997. *Environmental Scarcities, State Capacity, and Civil Violence: The Case of Indonesia*. Cambridge, Mass.: American Academy of Arts and Sciences.

Barber, C. V., N.C. Johnson, and E. Hafild. 1994. *Breaking the Logjam: Obstacles to Forest Policy Reform in Indonesia and the United States*. Washington, D.C.: World Resources Institute.

Bird, D. M. 1997. "National Guidelines on the Protection of Forests Against Fire." Consultancy report for the Integrated Forest Fire Management Project, presented at the International Workshop on National Guidelines on the Protection of Forests against Fire, Bogor, Indonesia, December 8–9.

Blakeney, J. 1998. *"Where There's Smoke, There's Fire!" Southeast Asia's Forest Fires of 1997/98*. World Bank Economic Development Institute Issues Paper. Washington, D.C. March.

Boehm, V. 1999. "Problems concerning the Million-Hectare Rice Project (PLG) and new Presidential Decree 80/1999." Unpublished Report. Kalteng Consultants, Hoehenkirchen, Munich, Germany. August 15.

Boer, C. 1989. "Investigations of the Steps Needed to Rehabilitate the Areas of East Kalimantan Seriously Affected by Fire: Effects of the Forest Fires of 1982/83 in East Kalimantan towards Wildlife." FR Report No. 7. Deutsche Forest Service/ITTO/GTZ, Samarinda, Indonesia.

Brauer, M. 1997. "Assessment of Health Implications of Haze in Malaysia. Mission Report to the World Health Organization Regional Office for the Western Pacific." November.

Brookfield, H., L. Potter, and Y. Byron. 1995. *In Place of the Forest: Environmental and Socioeconomic Transformation in Borneo and the Eastern Malay Peninsula*. Tokyo: United Nations University Press.

Brown, D.W. 1999. *Addicted to Rent: Corporate and Spatial Distribution of Forest Resources in Indonesia; Implications for Forest Sustainability and Government Policy*. Jakarta: Indonesia-UK Tropical Forest Management Programme Report No. PFM/EC/99/06. September.

Bruenig, E. F. 1996. *Conservation and Management of the Tropical Rain Forest: An Integrated Approach to Sustainability*. Cambridge, U.K.: CAB International/University Press.

Bryant, D., D. Neilsen, and L. Tangley. 1997. *The Last Frontier Forests: Ecosystems and Economies on the Edge*. Washington D.C.: World Resources Institute.

Byron, N., and G. Sheperd. 1998. "Indonesia and the 1997–98 El Niño: Fire Problems and Long-Term Solutions." *ODI Natural Resource Perspectives*, No. 29 (April): 1–4.

Carr, F. 1998. "Mamberamo Madness." *Inside Indonesia*, No. 55 (July-September).

CIC Consulting Group. *Study on Palm Oil Industry and Plantation in Indonesia, 1997*. Jakarta: PT Capricorn Indonesia Consult, Inc.

De Beer, J. H., and M. J. McDermott. 1996. *The Economic Value of Non-Timber Forest Products in Southeast Asia.* 2d ed., rev. Amsterdam: Netherlands Committee for IUCN (World Conservation Union).

Dennis, R. 1998. "A Review of Fire Projects in Indonesia." Draft. Center for International Forestry Research (CIFOR). July 30.

Department of Forestry and Estate Crops, Indonesia. 1999. "Pedoman Penataan Batas Hutan dan Pengukuhan" [Guidelines for forest boundary delineation and demarcation]. Draft No. 7. Jakarta. January.

Diemont, W. H., G. J. Nabuurs, J. O. Rieley, and H. D. Rijksen. 1997. "Climate Change and Management of Tropical Peatlands as a Carbon Reservoir." In J. O. Rieley and S. E. Page, eds., *Tropical Peatlands.* Cardigan, U.K.: Samara Publishing.

Doi, T. 1990. "Present status of large mammals in Kutai National Park after a large scale fire in East Kalimantan." In H. Tagawa and N. Wirawan (eds), *A Research on the Process of Earlier Recovery of Tropical Rain Forest After a Large Scale Fire in Kalimantan Timur, Indonesia.* Kagoshima University.

Dove, M. R. 1988. "Introduction: Traditional Culture and Development in Contemporary Indonesia." In M. R. Dove, ed., *The Real and Imagined Role of Culture in Development: Case Studies From Indonesia,* 1–37. Honolulu: University of Hawaii Press.

Dove, M. R. 1985. "The Agroecological Mythology of the Javanese and the Political Economy of Indonesia." *Indonesia* 39: 1–36.

EIA (Environmental Investigation Agency) and Telapak Indonesia. 1999. *The Final Cut: Illegal Logging in Indonesia's Orangutan Parks.* London and Jakarta.

Evers, P. 1995. "A Preliminary Analysis of Land Rights and Indigenous People in Indonesia." Draft Working Paper. World Bank, Jakarta.

Ferrari, L. 1997. "Haze Disaster Assignment: PM_{10} Instrument Measurement Training and Suspended Particle Concentration Assessment." Report to the World Health Organization (Indonesia) on a Short-Term Consultancy. Jakarta. November.

Filer, C., and N. Sekhran. 1998. *Loggers, Donors and Resource Owners: Policy That Works for Forests and People Series No. 2: Papua New Guinea.* Jakarta: Port Moresby: National Research Institute; London: International Institute for Environment and Development.

Forrester, G., and R. J. May, eds. 1999. *The Fall of Soeharto.* Published in association with the Research School of Pacific and Asian Studies, Australian National University. Singapore: Select Books.

Fuller, D., and M. Fulk. 1998. "An Assessment of Fire Distribution and Impacts during 1997 in Kalimantan, Indonesia Using Satellite Remote Sensing and Geographic Information Systems." Unpublished manuscript. World Wide Fund for Nature, Jakarta.

Gellert, P. K. 1998. "The Limits of Capacity: The Political Economy and Ecology of the Indonesian Timber Industry, 1967–1985." Unpublished Ph.D. diss. University of Wisconsin, Madison. UMI Dissertation Services.

Giesen, W. 1996. "Habitat Types and Their Management: Danau Sentarum Wildlife Reserve, West Kalimantan, Indonesia." Wetlands International Indonesia Programme/Directorate General of Forest Protection and Nature Conservation, Ministry of Forestry. Bogor, Indonesia.

Gillis, M. 1988. "Indonesia: Public Policies, Resource Management, and the Tropical Forest." In R. Repetto and M. Gillis, eds., *Public Policies and the Misuse of Forest Resources,* 43–104. Cambridge, U.K.: Cambridge University Press.

GOI (Government of Indonesia). 1991. *Indonesia Forestry Action Programme.* Vol. 2. Jakarta.

GOI (Government of Indonesia), Ministry of Forestry and Estate Crops. 1998. *Forestry Statistics of Indonesia 1996/97.* Jakarta.

GOI (Government of Indonesia) and IIED (International Institute for Environment and Development). 1985. *Forest Policies in Indonesia: The Sustainable Development of Forest Lands.* 3 vol. Jakarta.

Goldammer, J. G. 1997. "Forest Fire and Smoke Management and Policy Issues in Indonesia and Neighboring Countries of Southeast Asia." Paper presented at the International Workshop on National Guidelines on the Protection of Forests against Fire, Bogor, Indonesia, December 8–9.

Goldammer, J. G. 1998. "Land Use, Climate Variability and Fire in Southeast Asia: Impacts on Ecosystems and Atmosphere." Presented at the Asia-Pacific Regional Workshop on Transboundary Pollution, Singapore, May 27–28.

Goldammer, J. G., and C. Price. 1997. "Potential Impacts of Climate Change of Fire Regimes in the Tropics based on MAGICC and a GISS GCM-Derived Lightning Model." *Climatic Change* (in press).

Goldammer, J. G., and B. Siebert. 1990. "The Impact of Droughts and Forest Fires on Tropical Lowland Rain Forest of East Kalimantan." In J. G. Goldammer, ed., *Fire in the Tropical Biota: Ecosystem Processes and Global Challenges,* 11–31. Ecological Studies 84. Berlin-Heidelberg: Springer-Verlag.

Golden Hope Plantations Berhad. 1997. *The Zero-Burning Technique for Oil Palm Cultivation.* Kuala Lumpur, Malaysia.

Gonner, C. 1998. "Conflicts and Fire Causes in a Sub-District of Kutai, East Kalimantan, Indonesia." Unpublished report of the GTZ Sustainable Forest Management Project, Samarinda, East Kalimantan.

Grip, H. 1986. "A Short Description of the Experimental Watershed Study at Sipitang, Sabah." Paper presented at the Workshop on Hydrological Studies in Sabah, Kota Kinabalu, April 28.

GTZ (German Technical Cooperation). 1998. *Haze Guide. Information and Recommendations on How to Cope with Haze from Forest and Land Fires.* Version 3b. Jakarta. August.

Guiness, P. 1994. "Local Society and Culture." In H. Hill, ed., *Indonesia's New Order: The Dynamics of Socio-Economic Transformation,* 267-304. St. Leonards, Australia: Allen and Unwin.

Hansen, J. M. G. 1997. "Proposed Communications Systems to be Used in Forest and Land Fire Prevention and Suppression Activities." Paper presented at the International Workshop on National Guidelines on the Protection of Forests against Fire, Bogor, Indonesia, December 8–9.

Heil, A. 1998. "Air Pollution Caused by Large Scale Forest Fires in Indonesia, 1997." Unpublished report for the German Technical Cooperation (GTZ) project on Strengthening the Management Capacities of the Indonesian Forestry Ministry (SCMP) and the Integrated Forest Fire Management Project (IFFM). Jakarta.

Herman, T. 1998. "Million Hectare Swampland Project Visit Observations and Recommend-ations." World Bank, Jakarta. May 31.

Indonesia–U.K. Tropical Forest Management Programme. 1999. "A Draft Position Paper on Threats to Sustainable Forest Management in Indonesia: Roundwood Supply and Demand and Illegal Logging." Report No. PFM/EC/99/01. Jakarta.

Iskandar, M. B. 1997. "Health and Mortality." In G. W. Jones and T. H. Hull, eds., *Indonesia Assessment: Population and Human Resources.* Singapore: Institute of Southeast Asian Studies.

Johns, R. J. 1989. "The Influence of Drought on Tropical Rain Forest Vegetation in Papua New Guinea." *Mountain Research and Development* 9(3): 248–51.

Kartodihardjo, H. 1999. "Toward an Environmental Adjustment: Structural Barriers to Forest Development in Indonesia." Draft. World Resources Institute, Washington D.C.

Kartodihardjo, H., and A. Supriono, 1999. "The Impact of Sectoral Development on Natural Forest: The Case of Timber and Tree Crop Plantations in Indonesia." Draft report for the Center for International Forestry Research (CIFOR). Bogor, Indonesia.

Kershaw, A. P. 1994. "Pleistocene Vegetation of the Humid Tropics of Northeastern Queensland, Australia." *Paleogeography, Paleoclimatology, and Paleoecology* 109: 399–412.

LEI (Indonesian Ecolabeling Institute). 1998a. *Lembaga Ekolabel Indonesia dan Sertifikasi Pengelolaan Hutan Produksi Lestari* [The Indonesian Ecolabeling Institute and the certification of sustain-able management in production forests]. Jakarta. June.

LEI (Indonesian Ecolabeling Institute). 1998b. *Minutes of Meeting between YLEI Board of Trustees and FSC.* Jakarta.

Leighton, M. 1983. "The El Nino-Southern Oscillation Event in Southeast Asia: Effects of Drought and Fire in Tropical Forest in Eastern Borneo." Unpublished report, Department of Anthropology, Harvard University.

Leighton, M. and N. Wirawan, 1986. "Catastrophic drought and fire in Borneo tropical rain forest associated with the 1982-1983 El Nino Southern Oscillation event," in G.T. Prance (ed.), *Tropical Forests and the World Atmosphere,* pp. 75-102, AAAS Selected Symposium 101. American Association for the Advancement of Science, Washington, D.C.

Leitch, C. J., D. W. Flinn, and R. H. M. van de Graaff. 1983. "Erosion and Nutrient Loss Resulting from Ash Wednesday (February 1983) Wildfires: A Case Study." *Australian Forestry* 46: 173–180.

Lennertz, R., and K. F. Panzer. 1984. "Preliminary Assessment of the Drought and Forest Fire Damage in Kalimantan Timur." Report by DFS German Forest Inventory Service, Ltd., for German Agency for Technical Assistance (GTZ).

Liew, S. C., O. K. Lim, L. K. Kwoh, and H. Lim. 1998. "A Study of the 1997 Forest Fires in South East Asia Using SPOT Quicklook Mosiacs." Paper presented at the 1998 IEEE International Geoscience and Remote Sensing Symposium, Seattle, July 6–10.

Lilley, R. 1998. "Reptile and Amphibian Survey in Central Kalimantan with Special Reference to the Impacts of the Forest Fires." Unpublished report, WWF-Indonesia. No lizards were found in open burned peat swamp, but some snakes were found in nearby areas. Three land turtle species known to inhabit the area were not observed in burned or unburned areas, but frogs and tadpoles were observed in the survey area.

Mackie, C. 1984. "The Lessons behind East Kalimantan's Forest Fires." *Borneo Research Bulletin* 16: 63–74.

Mackie, J., and A. MacIntyre. 1994. "Politics." In H. Hill, ed., *Indonesia's New Order: The Dynamics of Socio-Economic Transformation,* 1–53. St. Leonards, Australia: Allen and Unwin.

Malingreau, J. P. 1987. "The 1982–83 Drought in Indonesia: Assessment and Monitoring." In M. Glantz, R. Katz, and M. Krenz, eds., *Climate Crisis: The Spatial Impacts Associated with the 1982-83 Worldwide Climate Anomalies,* 11– 18. Boulder, Colo.: National Center for Atmospheric Research; Nairobi: United Nations Environment Programme.

Malingreau, J. P., G. Stephens, and L. Fellows. 1985. "Remote Sensing of Forest Fires: Kalimantan and North Borneo in 1982–83." *Ambio* 14(6): 314–21.

Maltby, E. 1997. "Developing Guidelines for the Integrated Management and Sustainable Utilisation of Tropical Lowland Peatlands." In J. O. Rieley and S. E. Page, eds., *Tropical Peatlands.* Cardigan, U.K.: Samara Publishing.

Maltby, E., C. P. Immirzi, and R. J. Stafford, eds. 1996. *Tropical Peatlands of Southeast Asia.* Gland, Switzerland: IUCN (World Conservation Union).

Marten, G. G., ed. 1986. *Traditional Agriculture in Southeast Asia: A Human Ecology Perspective.* Boulder, Colo., and London: Westview Press.

Mayer, J.H. 1989. *Socioeconomic Aspects of the Forest Fire of 1982/83 and the Relation of Local Communities Towards Forestry and Forest Management.* Samarinda, Indonesia: FR Report No. 8, German Agency for Technical Cooperation (GTZ)/ITTO.

Michielsen, W. J. M. 1882. "Verslag eener reis door de boven districten der Sampit en Katingen riviersen in maart en april 1880" (Dutch). *Tijdschrift voor Indissche Taal, Land-en Volkenkunde* 28: 1–87.

Ministry of Forestry, Indonesia. 1996. *Indonesian Forestry Action Program.* Rev. vers. Jakarta.

Momberg, F., K. Atok, and M. Sirait. 1996. *Drawing on Local Knowledge: A Community Mapping Training Manual. Case Studies from Indonesia.* Jakarta: Ford Foundation, Yayasan Karya Sosial Pancur Kasih, and World Wide Fund for Nature Indonesia Programme.

National Institute of Health Research and Development, Ministry of Health, Indonesia. 1998. "A Cohort Study on Health Impacts of Haze from Forest Fires." Research design presented at the Bi-Regional Workshop on Health Impact of Haze-Related Air Pollution, Institute of Medical Research, Kuala Lumpur, Malaysia, June 1–4.

Natural Resources Institute, University of Greenwich (U.K.). 1996. "A Review of ODA Forestry Sector Strategy for Indonesia." Report prepared for the Overseas Development Administration. October. Jakarta.

Nicholls, N. 1993. "ENSO, Drought and Flooding in South-East Asia." In H. Brookfield and Y. Byron, eds., *South-East Asia's Environmental Future: The Search for Sustainability*, 154–74. Tokyo: United Nations University Press; Oxford, U.K.: Oxford University Press.

O'Brien, T.G., M.F. Kinaird, Sunarto, A.A. Dwiyahreni, W.H. Rombang and K. Anggraini, 1998. "Effects of the 1997 Fires on the Forest and Wildlife of the Bukit Barisan Selatan National Park, Sumatra." Unpublished report, Wildlife Conservation Society Indonesia Program and University of Indonesia.

Pangestu, M. 1989. "East Kalimantan: Beyond the Timber and Oil Boom." In H. Hill, ed., *Unity and Diversity: Regional Economic Development in Indonesia since 1970.* Singapore: Oxford University Press.

Peluso, N. L. 1992. *Rich Forests, Poor People: Resource Control and Resistance in Java.* Berkeley: University of California Press.

Peluso, N. L. 1995. "Whose Woods are These? Counter-Mapping Forest Territories in Kalimantan, Indonesia." *Antipode* 27(4): 383–406.

Phonboon, K., P. Kanatharana, O. Paisarnuchapong, and S. Agsorn. 1998. *Health and Environmental Impacts from the 1997 ASEAN Haze in Southern Thailand.* Bangkok: Health Systems Research Institute.

Pickford, S. G. 1995. "Fuel Types and Wildfire Hazard in the Bukit Soeharto Project Area." Final consultancy report for the Integrated Forest Fire Management Project. Samarinda, Indonesia. August.

Poole, P. 1995. *Indigenous Peoples, Mapping and Biodiversity Conservation.* Peoples and Forest Program Discussion Paper Series. Washington, D.C.: Biodiversity Support Program.

Poppele, J., S. Sumarto, and L. Pritchett. 1999. "Social Impacts of the Crisis: New Data and Policy Implications." Draft paper prepared for the World Bank. Jakarta.

Potter, L. 1988. "Indigenes and Colonisers: Dutch Forest Policy in South and East Borneo (Kalimantan) 1900 to 1950." In J. Dargavel, K. Dixon, and N. Semple, eds., *Changing Tropical Forests: Historical Perspectives on Today's Challenges in Asia, Australia and Oceania.* Canberra: Centre for Resource and Environmental Studies.

Potter, L., and J. Lee. 1998a. "Oil Palm in Indonesia: Its Role in Forest Conversion and the Fires of 1997/98." Report prepared for the World Wide Fund for Nature Indonesia Programme. Jakarta. October.

Potter, L., and J. Lee. 1998b. *Tree Planting in Indonesia: Trends, Impacts and Directions.* Final Report of a consultancy for the Center for International Forestry Research (CIFOR). Adelaide, Australia.

Radjagukguk, B. 1997. "Peat Soils of Indonesia: Location, Classification and Problems for Sustainability." In J. O. Rieley and S. E. Page, eds., *Tropical Peatlands.* Cardigan, U.K.: Samara Publishing.

Ramon, J., and D. Wall. 1998. "Fire and Smoke Occurrence in Relation to Vegetation and Land Use in South Sumatra Province, Indonesia with Particular Reference to 1997." Unpublished paper, European Commission, Forest Fire Prevention and Control Project. Jakarta.

Ridder, R. M. 1995. "Vegetation Inventory, Forest Classification and Fire Risk Mapping by Remote Sensing and GIS." Consultancy report for the Integrated Forest Fire Management Project. Samarinda, Indonesia. July.

Rieley, J. O. 1999. "Death of the Mega Rice Project! Creation of Another Monster?" Unpublished report. Kalimantan Tropical Peat Swamp Forest Research Project, School of Geography, University of Nottingham, UK. August. 22.

Rieley, J. O., and S. E. Page, eds. 1997. *Tropical Peatlands.* Cardigan, U.K.: Samara Publishing.

Romm, J. 1980. " Forest Development in Indonesia and the Productive Transformation of Capital." Presented at the Ninth Annual Conference on Indonesian Studies, Berkeley, Calif., July 31–August 3.

Sanderson, S., with S. Bird. 1998. "The New Politics of Protected Areas." In K. Brandon, K. H. Redford, and S. E. Sanderson, eds., *Parks in Peril: People, Politics, and Protected Areas*, 441–54. Washington, D.C.: Island Press.

Schindler, L., W. Thoma, and K. Panzer. 1989. *The Kalimantan Forest Fire of 1982–3 in East Kalimantan. Part I: The Fire, the Effects, the Damage and Technical Solutions*. FR Report No. 5. Jakarta: German Agency for Technical Cooperation (GTZ)/ITTO.

Schindler, L. 1998. "Fire Management in Indonesia: Quo Vadis?" Paper presented at the International Cross-Sectoral Forum on Forest Fire Management in South East Asia, Jakarta, December 8–9.

Schwarz, A. 1994. *A Nation in Waiting: Indonesia in the 1990s*. St. Leonards, Australia: Allen and Unwin.

Senior Management Advisory Team, Indonesia–U.K. Tropical Forest Management Programme. 1998. "Draft Response to Proposed IMF Forestry Sector Reforms." Report No. SMAT/98/EC/01. Jakarta.

Shimokawa, E. 1988. "Effects of fire of tropical forest on soil erosion," in H. Tagawa and N. Wirawan (eds), *Research on the Process of Earlier Recovery of Tropical Rain Forest After a Large Scale Fire in Kalimantan Timur, Indonesia* pp. 2-11. Occasional Paper No. 14, Research Center for the South Pacific, Kagoshima University.

Singapore. 1998. "Singapore Country Report on the 1997 Smoke Haze for the WHO Bi-Regional Workshop on the Health Impacts of Haze-Related Air Pollution," Kuala Lumpur, June 1–4.

Sowerby, J. and Yeager, C.P. 1997. "Fire effects on forests, forest wildlife and associated ecosystem processes." Unpublished manuscript, WWF-Indonesia Kayan Mentarang Project, East Kalimantan, Indonesia. October.

State Ministry for Environment, Indonesia. 1998. "Analisis Kebijakan: Penanggulangan Kebakaran Hutan dan Lahan Tahun 1997" [Policy analysis: response to the 1997 land and forest fires]. Jakarta. March.

State Ministry for Environment and UNDP (United Nations Development Programme). 1998. *Forest and Land Fires in Indonesia. Volume 1. Impacts, Factors and Evaluation*. Jakarta. September.

State Ministry for Environment, Indonesia. 1998. *Forest and Land Fires in Indonesia. Vol. 2: Plan of Action for Fire Disaster Management*. Jakarta.

Sunderlin, W. D. 1998. "Between Danger and Opportunity: Indonesia's Forests In an Era of Economic Crisis and Political Change." September 11, 1998. Available at http://www.cgiar.org/cifor.

Sunderlin, W. D., and I. A. P. Resosudarmo. 1996. *Rates and Causes of Deforestation in Indonesia: Towards a Resolution of the Ambiguities*. Occasional Paper No. 9. Bogor, Indonesia: Center for International Forestry Research.

Susilo, A. and J. Tangetasik, 1986. "Dampak kebakaran hutan terhadap perilaku orangutan (Pongo pygmaeus) di Taman Nasional Kutai" (Indonesian). Wanatrop 1(2).

Suzuki, A. 1988. "The socioecological study of orangutans and forest conditions after the big forest fire and drought, 1983 in Kutai National Park, Indonesia," in H. Tagawa and N. Wirawan (eds), *A Research on the Process of Earlier Recovery of Tropical Rain Forest After a Large Scale Fire in Kalimantan Timur, Indonesia*. Occasional Paper No. 14, Research Center for the South Pacific, Kagoshima University.

Tantra, I. G. M., H. Basuki, and A. Dwiyono. 1998. "Management of the Forest Conservation and Protected Areas of Indonesia." Presented at the International Forest Conservation and Protected Areas Expert Workshop, Canberra, Australia, September 8–10.

Tim Teknis Pengembangan Lahan Gambut Di Propinsi Kalimantan Tengah [Technical Team for Peatland Development in Central Kalimantan Province], 1997. "Analisis Mengenai Dampak Lingkungan Regional Pengembangan Lahan Gambut Satu Juta Hektar Di Propinsi Kalimantan Tengah" [Regional environmental impact analysis for the one million hectare peat land development in Central Kalimantan Province]. Jakarta: January.

Tomich, T. P., A. M. Fagi, H. de Foresta, G. Michon, D. Murdiyarso, F. Stolle, and M. van Noordwijk. 1998. "Indonesia's Fires: Smoke as a Problem, Smoke as a Symptom." *Agroforestry Today* 10(1, January-March): 4–7.

Trenberth, K. E., and T. J. Hoar. 1997. "El Niño and Climate Change." *Geophys. Res. Lett.* 24(23): 3057–60.

USDA (U.S. Department of Agriculture). 1997. "Health Hazards of Smoke: Recommendations of the Consensus Conference, April 1997." Washington, D.C. June. Cited in A. Heil, 1998, "Air Pollution Caused by Large Scale Forest Fires in Indonesia, 1997." Unpublished report for the German Technical Cooperation (GTZ) project on Strengthening the Management Capacities of the Indonesian Forestry Ministry (SCMP) and the Integrated Forest Fire Management Project (IFFM). Jakarta. July.

USEPA (U.S. Environmental Protection Agency). 1998. *Report on U.S. EPA Air Monitoring of Haze from S.E. Asia Biomass Fires*. EPA/600/R-98/071. Washington, D.C. June.

van Steenis, C. G. G. J. 1957. "Outline of the Vegetation Types in Indonesia and Some Adjacent Regions." *Eighth Pacific Science Congress*, Vol. I Botany, 61–97.

Vayda, A. P. 1998. "Finding Causes of the 1997–98 Indonesian Forest Fires: Problems and Possibilities." Report prepared for the World Wide Fund for Nature Indonesia Programme, Jakarta. Draft. December 23.

Vayda, A. P., C. J. Pierce Colfer, and M. Brotokusomo. 1980. *Interactions between People and Forests in East Kalimantan.* Honolulu: East-West Center Environment and Policy Institute.

Verstappen, H. T. 1980. "Quaternary Climatic Changes and Natural Environment in SE Asia." *GeoJournal* 4(1): 45–54.

Vidal, J. 1997. "The Thousand Mile Shroud: When the Earth Caught Fire." *The Guardian Weekend,* November 8.

Wakker, E. 1998. "Introducing Zero-Burning Techniques in Indonesia's Oil Palm Plantations." Report prepared for the World Wide Fund for Nature Indonesia Programme. Jakarta. April.

WALHI (Indonesian Foundation for the Environment). 1999. "Nasib Hutan Kita, Tidak di Tangan Rakyat: Studi WALHI Kebakaran Hutan dan Lahan 1997/1998" [The fate of our forests, not in the hands of the people: WALHI's study of the forest and land fires of 1997/1998]. Unpublished report. Jakarta.

Wells, M., and others. 1992. *People and Parks: Linking Protected Areas Management with Local Communities.* World Bank/World Wide Fund for Nature/U.S. Agency for International Development.

Whitmore, T. C. 1984. *Tropical Rain Forests of the Far East.* 2d ed. Oxford, U.K.: Clarendon Press/Oxford University Press.

Whitmore, T. C. 1990. *An Introduction to Tropical Rain Forests.* Oxford, U.K.: Clarendon Press.

Wirawan, N. 1985. *Kutai National Park Management Plan 1985-1990.* Bogor: WWF/IUCN.

Wirawan, N. 1993. "The Hazard of Fire." In H. Brookfield and Y. Byron, eds., *South-East Asia's Environmental Future: The Search for Sustainability,* 242–60. Tokyo: United Nations University Press; Oxford, U.K.: Oxford University Press.

World Bank. 1993. "Indonesia Forestry Sector Review." Draft. Jakarta. April.

World Bank. 1994a. "Indonesia. Environment and Development: Challenges for the Future." Report No. 12083-IND. Washington D.C.

World Bank 1994b. "Indonesia Transmigration Program: A Review of Five Bank-Supported Projects." Report No. 12988. Washington, D.C.

World Bank. 1995. "The Economics of Long Term Management of Indonesia's Natural Forests." Unpublished report. Jakarta.

World Bank. 1997. "Investing in Biodiversity: A Review of Indonesia's Integrated Conservation and Development Projects." Draft. Jakarta. June 10.

World Bank. 1998a. *Indonesia in Crisis: A Macroeconomic Update.* Washington, D.C.

World Bank. 1998b. "World Bank Involvement in Sector Adjustment for Forests in Indonesia: The Issues." Unpublished memorandum. Jakarta.

World Bank. 1999a. "Ensuring a Future for Indonesia's Forests (or Ensuring a Future Indonesian Forest)." Paper presented to the Consultative Group on Indonesia, Paris, July 29–30.

World Bank. 1999b. "Indonesia: From Crisis to Opportunity." Washington, D.C. July 21.

World Bank. 1999c. "Deforestation in Indonesia. A Preliminary View of the Situation in Sumatra and Kalimantan." Jakarta. Working Draft, August 16.

WWF-Indonesia and EEPSEA (Economy and Environment Programme for South East Asia). 1998. *The Indonesian Fires and Haze of 1997: The Economic Toll.* Singapore and Jakarta.

Yajima, T. 1988. "Change in the terrestrial invertebrate community structure in relation to a large fire at Kutai National Park, East Kalimantan (Borneo), Indonesia," in H. Tagawa and N. Wirawan (eds), *A Research in the Process of Earlier Recovery of Tropical Rain Forest After a Large Scale Fire in Kalimantan Timur, Indonesia,* Occasional Paper No. 14, Research Center for the South Pacific, Kagoshima University.

Yeager, C.P. 1998. "Interim Report on Fire Impacts on Tanjung Puting National Park." Unpublished report, World Wide Fund for Nature–Indonesia. August.

Yeager, C., and G. Fredriksson. 1999. "Draft Paper: Fire Impacts on Primate and Other Wildlife in Kalimantan, Indonesia during 1997/98." World Wide Fund for Nature Indonesia Programme. Jakarta.

Zakaria, R. Y. 1999. "Masalah Agraria & Kelembagaan Adat: Mari Menata Ulang Hubungan Raykat-Negara" [The agrarian problem and *adat* institutions: a call to reorder society-state relations]. Paper presented at the Discussion Dialogue on the Direction and Strategy for Agrarian Reform, Bogor, Indonesia, March 16–17.

Zerner, C. 1992. *Indigenous Forest-Dwelling Communities in Indonesia's Outer Islands: Livelihood, Rights, and Environmental Management Institutions in the Era of Industrial Forest Exploitation.* Consultancy Report prepared for the World Bank Indonesia Forestry Sector Policy Review. Washington, D.C.: Resource Planning Corporation.

Zuhud, E. A. M., and Haryanto, eds. 1994. *Pelestarian Pemanfaatan Keanekaragaman Tumbuhan Obat Hutan Tropika Indonesia* [Conservation and utilization of the diversity of medicinal plants of Indonesian tropical forests]. Bogor, Indonesia: Bogor Agricultural Institute and the Indonesian Tropical Institute (LATIN).

APPENDIX A
BAPPENAS–ADB METHODOLOGY FOR ESTIMATING ECONOMIC COSTS OF THE 1997–98 FIRES IN INDONESIA

FORESTS AND TIMBER

The total area burned was estimated at 9.8 million hectares (ha), 49 percent (4.65 million ha) of which is on forest land (forest of all categories but excluding plantations, both logged and unlogged). Estimates of timber destroyed were based on average standing volumes by island and forest type from the National Forest Inventory.[1] Estimates of the proportion of standing volume burned were based on a logging residue survey undertaken as part of the BAPPENAS-ADB project and an extensive field survey completed in a national park. From this work it was estimated that 30 percent of the basal area in burned areas was destroyed by fires.

The value of the timber destroyed was based on two estimates of the economic rent of the forest by island group, one based on current forest practices ($1.4 billion) and the other assuming reform of the log market ($2.1 billion).

Losses of trees below harvest age were also included by estimating reduction in volumes growing into the exploitable size classes, discounted into net present volumes, and given a value based on the two economic rent models noted above at $256 million and $377 million, respectively.

NON-TIMBER FOREST PRODUCTS

A socio-economic survey undertaken in East Kalimantan concluded that on average, rural households suffered a loss equivalent to $722 for the year following the fires. It was not possible to extrapolate these data to give the losses of non-timber forest products for area of forest burned because the socio-economic survey could not be linked to specific areas of forest. Surveys undertaken in a national park, however, which were linked to a specific and defined extensive area of forest, showed that total non-timber forest production was estimated at $28/ha/year in 1998 prices. Based on the assumptions derived for the degree of burned and lost trees, and the assumptions that non-timber forest production would gradually be re-established over a 20-year period, the aggregate loss of non-timber forest production from the fires was estimated at $586 million.

FLOOD PROTECTION, EROSION AND SILTATION

Based on a 1997 report covering 39 river catchments, which estimated the protection against flood damage afforded by forests at $91.60/ha/year, combined with the assumptions about loss of tree cover described above, the value of lost flood protection was estimated at $413 million.

The protection against erosion and siltation provided by forests was assigned a total value of $6,040/ha. This assumes that when forest is converted, it is lost forever. By calculating the discounted cashflow that would yield a net present value of $6,040, it is possible to estimate the protective function for the first few years. When this figure is multiplied by the area affected by the fires, an estimated loss due to erosion and siltation of $1.6 billion was determined.

CARBON EMISSIONS

The project team estimated that 757.5 million metric tons of carbon dioxide (CO_2) was produced during the 1997–98 fires (more than 75 percent of this as a result of the combustion of peat). Power generating companies that wish to offset their greenhouse gas emissions by either preventing the loss of carbon into the atmosphere or fixing atmospheric carbon through afforestation projects are prepared to pay between $6 and $8 per ton of carbon fixed or saved. Based on these assumptions, the total cost of carbon released into the atmosphere (based on $7/ton) was estimated at more than $1.4 billion. This figure is conservative; other estimates have put the amount of CO_2 produced at 3.7 billion tons, nearly five times the figure used here.

TIMBER PLANTATIONS

Estimated losses on timber plantations were calculated by assuming that the areas burned were evenly distributed over the different age classes. Based on field observations, it was determined that plantations less than 3 years old were completely destroyed but that plantations more than 3 years were only 30 percent destroyed. The figures for area burned were multiplied by the establishment costs compounded to present day terms to give an estimate of the loss in terms of establishment costs. Estimates of profit foregone were also included. In this manner, the total loss of timber plantation value was estimated at $94 million.

ESTATE CROPS

Official area estimates of estate crops destroyed tallied with the estimates of the BAPPENAS–ADB study, and thus, the figure of $319 million determined by the Ministry of Environment valuation study[2] was used.

AGRICULTURE

Agricultural losses incurred during 1997 and 1998 were due to drought as well as fires and haze. By analyzing past trends of agricultural production it was possible to predict the level of production for 1997 and 1998 had there been no drought, fires, or haze. Estimates of lost production were derived by subtracting the actual production from the predicted production. Findings showed that rice production had significantly decreased (beyond normal variability) by 2.6 million tons in 1997 and 7 million tons in 1998. The economic cost is the expense of trying to grow the crop (i.e., wasted seed, fertilizer, pesticide, labor, etc.) plus the profit foregone by the farmers—the equivalent of the farm gate price. The total economic cost of lost rice production was estimated at $1.9 billion, to which was added the net cost of importing rice as a substitute, for a total estimated agricultural loss of $2.4 billion.

HEALTH

Official statistics for the health impacts of the haze in 1997 were given in the Ministry of Environment valuation study. The 1998 smoke and haze event covered three provinces in Kalimantan for roughly the same period and to the same intensity as in 1997. It was therefore assumed that health impacts incurred in these provinces in 1998 would be the same as in 1997. The impacts were then multiplied for standard health care costs and estimates for lost productivity to give a total health impact estimate of $145 million.

TOURISM

Tourism, an important economic sector for Indonesia, declined significantly in 1997 and 1998, but not all of this reduction can be attributed to the fires; other factors such as the Asian economic crisis and the political unrest in 1998 also contributed. By analyzing the trends of tourist arrivals by region of origin, it is possible to predict the numbers of tourists that would have arrived had these events not occurred. Subtracting the actual arrivals from the numbers predicted by the trends produced an estimate of the loss in the numbers of tourist arrivals. Assuming standard profit margins and overheads it was then possible to estimate the economic loss in tourism due to the fires and haze, which was determined to be $111 million (cf. the WWF-EEPSEA estimate of $70 million for 1997 alone.)[3]

OTHER LOSSES

The officially reported losses for damage in transmigration areas, transport losses, and firefighting costs have also been included, totalling $46 million for these three items.

Source:
National Development Planning Agency (BAPPENAS), 1999. *Final Report, Annex I: Causes, Extent, Impact and Costs of 1997/98 Fires and Drought.* Asian Development Bank (ADB) Technical Assistance Grant TA 2999-INO, Planning for Fire Prevention and Drought Management Project (April).

Notes:
1. Ministry of Forestry, 1996. *Final Forest Resources Statistics Report.* Jakarta.
2. State Ministry for Environment and UNDP, 1998.
3. WWF Indonesia Programme and EEPSEA, 1998.

The ecological impacts of the 1997–98 fires have yet to be systematically assessed in the field, except for the preliminary studies discussed below. Considerable prior information exists, however, about the ecological impacts of forest fires on tropical forest ecosystems generally—including a great deal of data concerning the East Kalimantan fires of 1982–83; these data are used to extrapolate probable impacts.

FOREST VEGETATION

The effects of fire on the vegetation in forest ecosystems are complex, varying with the type of forest, degree and recentness of disturbance, level of drought, and incidence of repeated fire episodes.[1] Lowland rainforests and peat swamp forests, for example, two forest types particularly affected by the 1997–98 fires, react very differently to fire.

The immediate effect of a forest fire is to reduce vegetation to nutrient-rich ash, which can nourish the beginnings of a new forest. However, if the fire is very hot, the soil surface hardens, making it difficult for seeds to sprout, and causing the ash to be washed away by the first heavy rain.[2] Intense burns and subsequent soil erosion result in the loss of other soil constituents that facilitate vegetation regrowth, such as organic matter, soil organisms that accelerate plant matter decomposition, and specialized fungi that assist key tree species to absorb nutrients. A comparison of soil erosion rates between burned and unburned forest in Kutai National Park after the 1982–83 fires showed that erosion had accelerated more than tenfold in the burned areas.[3] Soil erosion does not occur in the aftermath of peat swamp fires, but ash and other fire residues are washed away and the surface level of burned peat is lowered by combustion losses.

As noted above in the discussion of the 1982–83 fires, improperly logged forests are particularly fire prone because excessive amounts of waste wood are left on the forest floor and the forest canopy is opened, causing ground vegetation and dead branches to dry out quickly. Heavily disturbed forest tends to burn almost completely, leaving few live trees. Pristine forest is much less likely to burn, and when it does, usually only ground-level vegetation is consumed leaving the middle and upper tree layers intact. Lightly-burned pristine forest is quick to recover after a fire. Moderately to heavily burned forests take decades or centuries to regenerate due to an invasion of pioneer tree species and the loss of seeds and seedlings of species normally found in a mature forest. Heavily burned forest may be converted to grasslands by repeated intentional burning. The primary vegetation ecology question to be answered is whether repeated large-scale fires will upset the stability of forest ecosystems beyond the point of recovery.

Peat swamp forests present a special case, because they are particularly vulnerable to fire and produce the most noxious smog of any forest type when they burn. A significant portion of the haze in 1997 was generated by peat fires, which are quite different from fires in lowland forest. Peat fires typically burn underground as well as above, produce relatively low heat, generate large amounts of smoke, eliminate the seedbank, and destroy the soil, which can take thousands of years to replace.[4]

Research carried out in peat forests affected by the 1997 fires at Central Kalimantan's Tanjung Puting National Park provides an indication of the impacts of fire on peat swamp forest vegetation. (*See Table B-1.*) Areas that burned in 1991 but were allowed to regenerate without disturbance until 1998 showed signs of rapid regeneration, whereas areas burned repeatedly with only short intervals between the fires showed much lower stem densities and species diversity.[5]

WILDLIFE

It is difficult to document the effects of fire on rain forest animals and insects because their populations can vary seasonally or in multi-year cycles, and the ecology of many species has not been well studied. Wildlife may be killed directly by the heat and smoke of fires or may subsequently weaken and die from lack of food and water or habitat loss.[6] Small, slow-moving animals are most likely to be killed outright by fires, and animals with highly specific food, habitat, shelter, or climate requirements are most at risk during the immediate post-fire period. Some larger creatures are capable of moving to other areas to escape fire but often stray into territory settled by humans and are captured or killed, as happened to many orangutans in Kalimantan during 1997. The loss of key organisms in ecosystems, such as pollinators and decomposers, can significantly slow the recovery of the forest ecosystem.[7] The changing composition of vegetation in a recovering forest may provide alternate or even superior food sources for some omnivores, generalist herbivores and insects, sometimes leading to dramatic increases in their populations after fires, and thereby changing the faunal composition of the forest.

The fauna of East Kalimantan's Kutai National Park received the most concentrated scientific attention immediately after the 1982–83 fires, and in the half decade that followed. These studies showed that most large mammals were still in the area, with wild pigs (*Sus spp.*) becoming abundant by taking advantage of new food sources,[8] and Banteng (*Bos javanicus*) still common because these large wild cattle also adapted their diets.[9] Many sambar deer (*Cervus unicolor*) and barking deer (*Muntiacus muntjak*) perished in the fires,[10] but their populations appeared to have recovered a year later.[11] The Malayan Sun Bear (*Helarctos malayanus*) is thought to have declined in Kutai, however, possibly beyond the point of recovery, while small carnivores like the Malay civet (*Vicerra tangalunga*), otter (*Lutra spp.*), leopard cat (*Felis bengalensis*), and flat headed cat (*Felis planiceps*) are thought to have increased in numbers in the years after the fires in response to an increase in prey animals.[12]

The 1982–83 fires caused high mortality among reptiles and amphibians,[13] and had a negative impact on swamp dwelling reptiles, but most, with the exception of the crocodile, eventually reappeared in their former habitat.[14] Snakes also reappeared in the forest, with the exception of large species such as pythons (*Phyton* spp.). In 1997, forest lizards were completely absent from burned areas of a national park in Sumatra one month after it burned.[15] A rapid survey in a peat swamp area of Central Kalimantan immediately after the 1997 fires yielded preliminary conclusions that most land and arboreal reptiles had probably died from the heat of the fires, whereas crocodiles, water turtles, and other species inhabiting relatively deep water had largely survived.[16]

The charismatic and relatively well-studied orangutan (*Pongo pygmaeus*) was the animal species that received the most media attention during both the 1982–83 and 1997–98 fires. Researchers in Kutai found that while some orangutans perished in the fires, and others were displaced and malnourished in its immediate aftermath, these omnivorous great apes were able to switch to eating bark and young stems until fruit reappeared in the forest.[17] Susilo and Tangketasik concluded that maturing secondary forest offers more food value to orangutans than primary forest. As early as 1984, orangutans were observed carrying infants born after the fires, proving the species' ability to adapt to the new conditions.[18]

The scale of the 1997–98 fires, however, exceeded the orangutan's ability to adapt to stressful situations. Hundreds of adults were killed by villagers in Central and East Kalimantan as they fled from the forest to escape the effects of the drought, smoke and fires. Immediately after the 1997 fires, a WWF-Indonesia researcher encountered 14 live orangutans in and around Tanjung Puting National Park, an area with a large and well-studied population of orangutans. A farmer outside the park told the researcher that he had killed a large male orangutan with a spear as it ate pineapples in the farmer's field. Many orphaned juveniles were sold for the pet trade, but as of April, 161 mostly young orangutans were in the care of the Semboja Orangutan Introduction Center in East Kalimantan. Even this safe haven was threatened by fire and food scarcity in the area. Primatologists believe that the 1997–98 fires will mark the beginning of a steeper downward trend in the already declining population of Bornean orangutans.

Mean Stem Density and Species Richness of Forest Vegetation in Relation to Fire Exposure at Tanjung Puting National Park, Central Kalimantan

	STEMS/HECTARE	SPECIES/HECTARE
Burned in 1994 and 1997	165	11
Burned repeatedly, including 1997	113	7
Burned in 1991	235	24
Unburned	951	35

Other primates fared relatively well after the 1982–83 fires because they adapted their diets to replace figs and other favored fruits from trees that had suffered high levels of damage and mortality. Leighton reported that both pig tailed macaques (*Macacca nemestrina*) and gibbons (*Hylobates muelleri*) in Kutai National Park took advantage of explosions in the populations of wood boring insects immediately after the fires. He detected no change in the behavior or activity of two gibbon families that he had studied prior to the fires.[19] Leaf eating monkeys (*Presbytis spp.*) were very difficult to find after the fires, and even six years later their densities were low. Proboscis monkeys *(Nasalis larvatus)* maintained their populations in mangrove forest (Boer, 1989), a vegetation type not heavily damaged by the fires. Two primitive primates, however, western tarsiers (*Tarcius bancanus*) and slow loris (*Nycticebus coucany*), were extinct or extremely reduced in number by 1986.[20] Seven years after the fires, natural succession favored figs, lianas, and other important primate fruit species,[21] boding well for the recovery of most primate populations.

Proboscis monkeys are a threatened species found almost exclusively in riverine and coastal habitats. Because riverine forest was heavily affected by the 1997–98 fires, this species has probably lost the greatest percentage of its remaining habitat of any primate species in Borneo.

Bird populations were reduced by direct impacts of the 1982–83 fires. During the fires, birds were observed by local people to become disoriented in the heavy smoke and fall to the ground.[22] Fruit-eating birds, especially hornbills, could not be found in Kutai immediately after the 1982–83 fires, presumably because the fires had killed a large percentage of the fruit trees upon which they depend for food.[23] Insect-feeding birds enjoyed an abundance of food in the aftermath of fires because wood-eating insect populations exploded in response to the enormous supply of dead wood.

Because of the short life cycles of most insects and other invertebrates, their populations react relatively quickly to ecological changes in the aftermath of fires. As mentioned above, the 1982–83 fires caused populations of wood-eating insects to increase dramatically, and butterfly populations exploded as a result of the abundance of nutrient-rich ash.[24] One entomologist studying the effects of the 1982–83 fires on invertebrate ecology in Kutai National Park found that the diversity and numbers of soil and litter dwelling invertebrates declined somewhat after the fire, but recovered within three years.[25] The rich, but little studied invertebrate fauna living in the canopies of tropical trees are presumably destroyed along with their habitat. The recovery of arboreal invertebrate diversity probably parallels forest regeneration.

There appear to be no scientific reports describing the effects on coral reefs and other nearshore marine ecosystems when the heavy rains of late 1983 flushed fire-generated sediment and polluted water into the Makassar Straight. This coastline is largely muddy due to the normal discharge of sediment from several large rivers, but offshore islands are ringed by coral reefs. Corals are quickly smothered and killed by even brief exposure to heavy sediment, and unusually high levels of fresh water discharge from rivers can kill corals by reducing the salinity of ocean water. Coastal ecology can also be negatively affected because the important fish nursery function of mangrove ecosystems may be impaired by excessive deposition of sediment. La Niña brought abnormally heavy rainfall towards the end of 1998, but even a normal rainy season would have serious consequences for marine ecosystems as the remains of rain forests are washed out to sea.

Notes:

1. Sowerby and Yeager, 1997.

2. Wirawan, 1993.

3. Shimokawa, 1988.

4. Yeager, 1998.

5. Ibid.

6. Boer, 1989.

7. Yeager, 1998.

8. Doi, 1990.

9. Wirawan, 1985.

10. Boer, 1989.

11. Mayer, 1989.

12. Doi, 1990.

13. Leighton and Wirawan, 1986.

14. Boer, 1989.

15. O'Brien and others, 1998.

16. Lilley, 1998.

17. Suzuki, 1988.

18. Susilo and Tangetasik, 1986.

19. Leighton, 1983.

20. Boer, 1989.

21. Schindler, Thoma and Panzer, 1989.

22. Mayer, 1989.

23. Leighton, 1983.

24. Wirawan, 1993.

25. Yajima, 1988.

APPENDIX C
CERTIFICATION PROCEDURES OF THE INDONESIAN ECOLABELING INSTITUTE (LEI)

Preliminary evaluation. A firm seeking certification of one of its units submits an application along with the documents and information specified in LEI's guidelines. If the documents are complete, the firm signs an agreement with LEI and pays the fees for the initial assessment of information contained in the documents. This assessment is carried out by Expert Panel I, which is appointed by LEI. Applicants are given a chance to explain and elaborate on their documentation before the panel. If Panel I agrees that the management unit is a valid candidate for certification, it recommends implementation of a field assessment and notes areas to which the field assessment should pay particular attention.

Field assessments. LEI itself does not carry out field assessments but, rather, certifies assessors who meet its published criteria. The applicant chooses the assessor from among those meeting the criteria through a process of open bidding. The winning bid is scrutinized by LEI, which, if there are no problems, issues a "no objections" letter. The applicant then concludes a contract for services with the assessor for the field assessment. Assessors may be private firms or NGOs, as long as they meet the requisite LEI criteria. LEI also designed the standardized curriculum and training manuals for field assessors. The first round of field assessor training, for 86 professionals, was held in June 1998.

The field assessment team carries out its assessment on the basis of LEI's set procedures, criteria, and indicators, complemented by the special concerns raised by Expert Panel I. The applicant is required to send one or more staff to accompany the team and guide it in the field. The accompanying staff must possess adequate knowledge and authority to directly clarify matters for the field assessment team as necessary. In parallel with the field assessment, LEI provides public notice (through local newspapers) to all stakeholders in the area that the management unit is under assessment and invites stakeholders to contribute additional views or information. LEI also encourages local NGOs, communities and other stakeholders to form a regional forestry consultative forum to facilitate the articulation of local concerns relating to the certification process.

Within 30 days of completing the field assessment, the assessor writes a report according to set guidelines and provides it to LEI.

Performance evaluation. After the applicant makes payment to LEI to cover the costs of this stage of the process, LEI forms and briefs Expert Panel II, which evaluates the management unit's performance on the basis of the assessor's field report and additional information that other stakeholders may have provided to LEI. Membership of this panel is the same as for Panel I, with the addition of four experts on the region where the operation in question is located. In addition to studying the report, the panel is given an opportunity to directly question the field assessors. The panel then ranks the performance of the applicant's management unit (using as grades the terms gold, silver, bronze, copper, or zinc) and makes its recommendation to LEI. Only those applicants attaining gold, silver, or bronze rankings are eligible for certification. The panel also makes recommendations to the applicant on actions it should take to bring its operation more into line with the LEI criteria for sustainable management.

The certification decision. LEI then issues a five-year certification, which is announced in the mass media. This decision is final, although an appeals process is provided under which any party can appeal the decision. (The firm in East Kalimantan that failed its assessment due to fires in its concession area has appealed the decision, giving the appeals process its first real test.) A maximum of three field audits may be carried out by LEI during the five-year period, but the first one must be carried out within the first two years. These audits can result in upgrading, downgrading, or revocation of the firm's certification.

MAP 1: INDONESIA'S REMAINING FRONTIER FORESTS

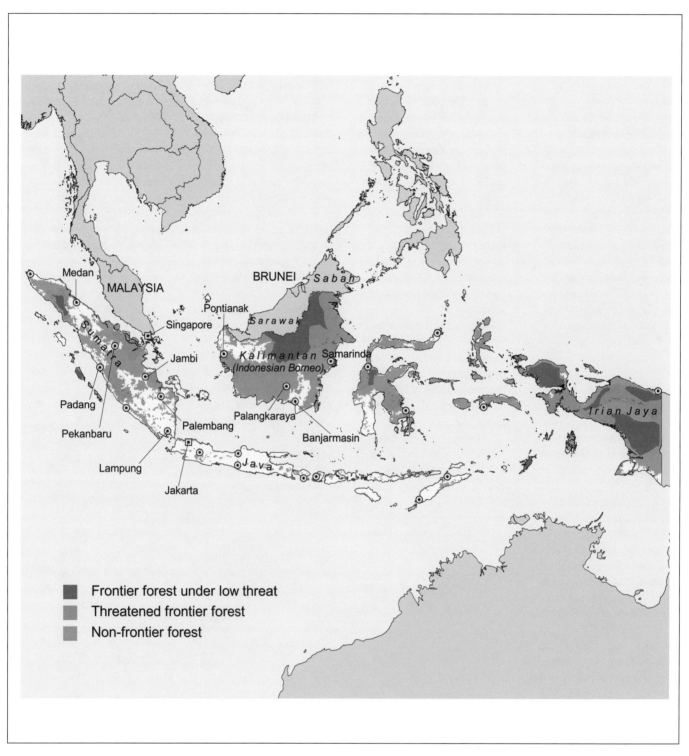

Frontier forest under low threat
Threatened frontier forest
Non-frontier forest

Source: Bryant, Nielsen, and Tangley, 1997.

Notes: (a) "Frontier forest" refers to large, ecologically intact and relatively undisturbed natural forests. "Non-frontier forests" are dominated by secondary forests, plantations, degraded forest, and patches of primary forest not large enough to qualify as frontier forest. "Threatened frontier forests" are forests where ongoing or planned human activities will eventually degrade the ecosystem. See Bryant, Nielsen, and Tangley for detailed definitions.

(b) This map was completed prior to the 1999 release of the results of a World Bank-assisted forest mapping effort that concluded that deforestation rates since 1986 have been 50 percent greater than hitherto assumed. Actual forest cover is therefore probably less than shown on this map.

MAP 2: DISTRIBUTION AND INTENSITY OF "HAZE" FROM FOREST FIRES IN INDONESIA, SEPTEMBER TO NOVEMBER 1997

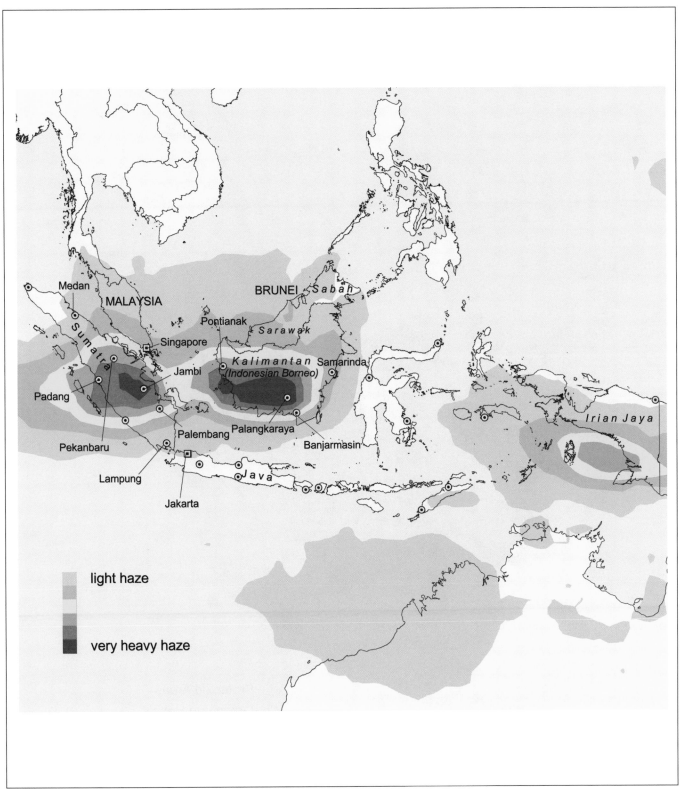

Source: Map composition by Y. Martin Hardiono, Telapak Indonesia, reproduced at World Resources Institute, 1999.

Notes: Haze distribution is the mean of cumulative haze distribution from September to November 1997. Derived from Earth Probe satellite data available on the NASA Total Ozone Monitoring System site at: http://jwocky.gsfc.nasa.gov/index.html

MAP 3: DISTRIBUTION OF WILD ORANGUTAN POPULATION, ACCUMULATED HOT SPOTS, AND PROTECTED AREAS IN KALIMANTAN, 1997-98

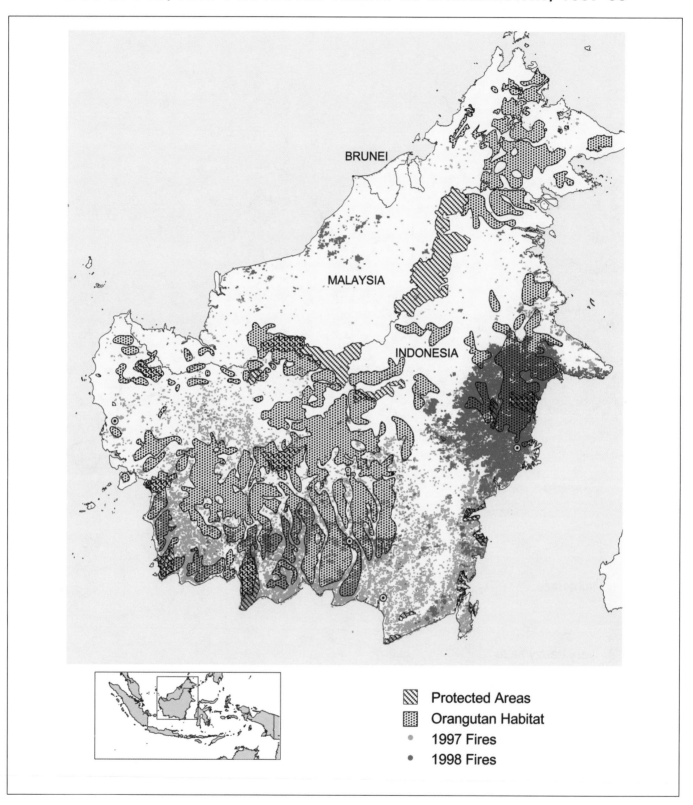

BRUNEI

MALAYSIA

INDONESIA

Protected Areas
Orangutan Habitat
1997 Fires
1998 Fires

Source: WWF–Indonesia, 1999. Orangutan Action Plan. Jakarta.

MAP 4: FOREST USES AND AREAS BURNED IN 1997-98, EAST KALIMANTAN PROVINCE

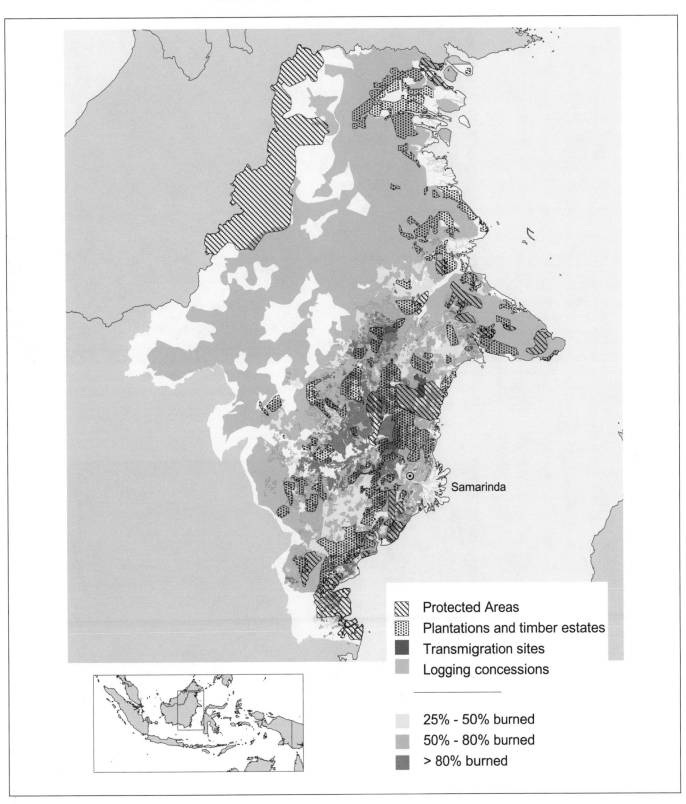

Samarinda

Protected Areas
Plantations and timber estates
Transmigration sites
Logging concessions

25% - 50% burned
50% - 80% burned
> 80% burned

Source: Map composition by Y. Martin Hardiono, Telapak Indonesia, reproduced at World Resources Institute, 1999.
Note: Fires data from German Technical Cooperation (GTZ) Integrated Forest Fires Management Project, Samarinda, Indonesia.

MAPS 5A-5C: LAND CLEARING AND FIRE ON THE CENTRAL KALIMANTAN MILLION-HECTARE RICE PROJECT. MAY 1995-JULY 1997

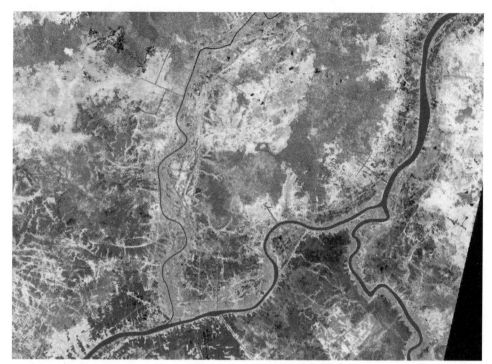

LANDSAT-image (30 km x 42 km) of the Dadahup area of the Million-Hectare Rice Project, Central Kalimantan, May 10, 1996. Green areas are peat swamp forest.

Source: Kalteng Consultants, Hoehenkirchen, Germany

The Dadahup area on May 29, 1997. Clearcutting of the peat swamp forest and construction of new canals are shown in red.

Source: Kalteng Consultants, Hoehenkirchen, Germany

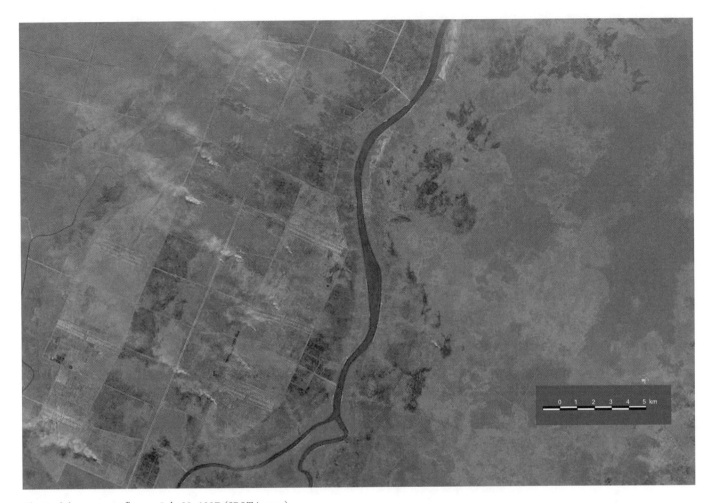

The Dadahup area in flames, July 29, 1997 (SPOT image).

Source: Kalteng Consultants, Hoehenkirchen, Germany

BOARD OF DIRECTORS

WORLD RESOURCES INSTITUTE

The World Resources Institute (WRI) is an independent center for policy research and technical assistance on global environmental and development issues. WRI's mission is to move human society to live in ways that protect Earth's environment and its capacity to provide for the needs and aspirations of current and future generations.

Because people are inspired by ideas, empowered by knowledge, and moved to change by greater understanding, the Institute provide—and helps other institutions provide—objective information and practical proposals for policy and institutional change that will foster environmentally sound, socially equitable development. WRI's particular concerns are with globally significant environmental problems and their interaction with economic development and social equity at all levels.

The Institute's current areas of work include economics, forests, biodiversity, climate change, energy, sustainable agriculture, resource and environmental information, trade, technology, national strategies for environmental and resource management, and business liaison.

In all of its policy research and work with institutions, WRI tries to build bridges between ideas and action, meshing the insights of scientific research, economic and institutional analyses, and practical experience with the need for open and participatory decision-making.

WORLD RESOURCES INSTITUTE
10 G Street, N.E.
Washington, D.C. 20002, USA
http://www.wri.org/wri

WORLD WIDE FUND FOR NATURE

The World Wide Fund for Nature (WWF) is one of the world's largest and most respected private conservation organizations. Based in Gland, Switzerland, WWF has a worldwide network of 27 national organizations, 5 associated organizations and 21 program offices, with over 4.7 million supporters worldwide. WWF has been active in Indonesia since the 1960s, and currently runs more than twenty ongoing projects at different field locations throughout Indonesia. In September 1996, the WWF Indonesia Foundation was established as a step towards becoming a WWF National Organization, resulting in a change of name in July 1998 from the WWF Indonesia Programme to WWF-Indonesia. In the early years, WWF's mission in Indonesia was primarily to preserve endangered wildlife. But its current mission has expanded, and now embraces preservation of biological diversity, sustainable use of natural resources, and reduced consumption and pollution. WWF-Indonesia's national office is located in Jakarta, with additional offices in Bali, East Kalimantan, and Irian Jaya, each focusing on one of Indonesia's major bioregions and working closely with local governments, NGOs, and communities.

TELAPAK INDONESIA FOUNDATION

Telapak is an Indonesian nongovernmental organization (NGO) based in Bogor, West Java. Founded in 1997, Telapak's objective is to support and strengthen sustainable and equitable management of Indonesia's forest and marine ecosystems and resources. Telapak works through field investigations, policy analyses, and the provision of information to policymakers, the media, and other NGOs. Its primary focus is on exposing policies and practices of government agencies, the private sector, and international financial institutions that are prejudicial to Indonesia's living environment and the interests of future generations, and proposing alternative policies for sustainable and equitable development. Telapak is strongly committed to working with like-minded individuals and organizations at the grassroots level throughout Indonesia, and therefore directs a considerable amount of energy to helping empower local NGOs and communities to serve as strong defenders of living natural resources in their areas. Telapak is the host institution for Forest Watch Indonesia, an independent forest monitoring network that is affiliated with Global Forest Watch, an international initiative hosted by the World Resources Institute.

THE WORLD RESOURCES INSTITUTE FOREST FRONTIERS INITIATIVE

The World Resources Institute Forest Frontiers Initiative (FFI) is a multi-disciplinary effort to promote stewardship in and around the world's last major frontier forests by influencing investment, policy, and public opinion. The FFI team is working with governments, citizens' groups, and the private sector in Amazonia, Central Africa, Indonesia, North America, and Russia. We also take part in pressing international discussions on forest policy.

We are motivated by the belief that there is a responsible way to use forests. We also see growing interest in finding alternatives to forest destruction that take advantage of the full economic potential of forests, not just immediate revenue from logging and forest clearing.

For each frontier forest region, FFI builds a network of policy-makers, activists, investors, and researchers to promote policy reform. Efforts to minimize the negative impacts of road-building and forest-clearing for agriculture and to stop illegal logging are part of this work.

In collaboration with a variety of partners, WRI is creating Global Forest Watch—an independent, decentralized, global forest monitoring network—which will facilitate the collection of all relevant information on forests and how they are being used as well as provide mechanisms for making this information available to anyone with a stake in the forest.

Business has a leading role to play. WRI is working with the forest products industry and others to create greater production and demand for goods from well-managed forests. We are developing case studies with innovative firms to demonstrate to others the business impacts and opportunities that sustainability presents.

To get access to information about FFI findings and activities and to find out how to participate, visit our website at **http://www.wri.org/wri/ffi/** or write to:

Forest Frontiers Initiative
World Resources Institute
10 G Street, N.E.
Washington, D.C. 20002, U.S.A.
Telephone: 202/729-7600
Fax: 202/729-7610
Email: ffi@wri.org